About Island Press

Since 1984, the nonprofit organization Island Press has been stimulating, shaping, and communicating ideas that are essential for solving environmental problems worldwide. With more than 1,000 titles in print and some 30 new releases each year, we are the nation's leading publisher on environmental issues. We identify innovative thinkers and emerging trends in the environmental field. We work with world-renowned experts and authors to develop cross-disciplinary solutions to environmental challenges.

Island Press designs and executes educational campaigns in conjunction with our authors to communicate their critical messages in print, in person, and online using the latest technologies, innovative programs, and the media. Our goal is to reach targeted audiences—scientists, policymakers, environmental advocates, urban planners, the media, and concerned citizens—with information that can be used to create the framework for long-term ecological health and human well-being.

Island Press gratefully acknowledges major support of our work by The Agua Fund, The Andrew W. Mellon Foundation, The Bobolink Foundation, The Curtis and Edith Munson Foundation, Forrest C. and Frances H. Lattner Foundation, The JPB Foundation, The Kresge Foundation, The Oram Foundation, Inc., The Overbrook Foundation, The S.D. Bechtel, Jr. Foundation, The Summit Charitable Foundation, Inc., and many other generous supporters.

The opinions expressed in this book are those of the author(s) and do not necessarily reflect the views of our supporters.

Water is for Fighting Over

Water is for Fighting Over

AND OTHER MYTHS ABOUT WATER IN THE WEST

John Fleck

Washington | Covelo | London

ISLAND PRESS is a trademark of the Center for Resource Economics.

Library of Congress Control Number: 2016938029

Printed on recycled, acid-free paper

Manufactured in the United States of America
10 9 8 7 6 5 4 3 2 1

Keywords: Colorado River Basin, Lake Mead, Morelos Dam, Minute 319, San Luis Río Colorado, cienega, agriculture-to-urban transfers, Metropolitan Water District of Southern California, Imperial Irrigation District, All-American Canal

To Lissa, from the headwaters to the sea

Whiskey's for drinkin', water's for fightin' over.
— apparently not Mark Twain

Contents

Acknowledgments

From the day Karl Flessa sat down with me in an Albuquerque bakery named Isabella's in the summer of 2009 and explained what happens when the Colorado River fails to reach the sea, the members of the river community have been unfailingly smart and generous. That day Karl embodied what I came to see as the two essential traits of this group I have come to know and love: a clear-eyed view of the difficulty of the Colorado's problems, and persistent optimism that they can be solved.

On the farmlands of the greater Colorado River Basin, Corky Herkenhoff, Tom Davis, Mark Smith, Bart Fisher, the Sharp brothers (Clyde and David), Pat Morgan, Jose Ramirez, and Ron Derma showed me how you move water through a desert to grow food. Tina Shields showed endless patience explaining water in the Imperial Valley and the importance of the Salton Sea. In Las Vegas, Kurtis Hyde taught me how to garden in a desert. In Phoenix, Kathryn Sorensen showed what it takes to keep the taps running for 1.5 million people who chose to build an improbable city in the desert. Bill Hasencamp helped me understand how Southern California has learned to live with less. John

Stomp showed me early on why water managers have to take a longer view into the future than most of us.

Tanya Trujillo first explained the strangeness of the Law of the River during a memorable meeting on the *Albuquerque Journal*'s patio, and then kept explaining it, again, and again, and again. Tom McCann's deep understanding of the history of the river's problems and his tenacious pursuit of solutions are a model. Jennifer Pitt and Mike Cohen demonstrated that there is no substitute for putting in the effort to understand how the system works, then spent countless hours helping me up the learning curve.

Mike Connor and the river's federal management community, past and present, were generous beyond measure with their time, insights, and data—Terry Fulp, Jennifer McCloskey, Bob Snow, Anne Castle, Carly Jerla, Dan Bunk, Rose Davis, Paul Miller, Joe Donnelly, and many more.

Doug Kenney's and Larry MacDonnell's scholarship have been central to my understanding of the river's law, politics, and history. Brad Udall's insights, friendship, and counsel have been crucial.

More than anyone, John Entsminger helped me understand the importance of "the network"—how a community of trust and reciprocity across difficult geographic and institutional boundaries is the only way to solve these problems.

Stanford University's Bill Lane Center for the American West and its then-director Jon Christensen gave important early support and encouragement for the project that became this book, and Jon egged me on when I needed it. The Water Education Foundation opened a key door.

My former employer, the *Albuquerque Journal*, especially Charlie Moore and Isabel Sanchez, long encouraged my journalistic obsession with water. When I needed to devote my full energy to this project, Bob Berrens of the University of New Mexico's Water Resources Program gave me a new home to think and write, and more importantly shared countless afternoon hours, in my office and his, hashing out the ideas that you see here. Bruce Thomson taught me about water and then literally gave me his old UNM office to write the book in. Melinda Harm Benson in UNM's Department of Geography and Environmental Studies helped me think through the crucial theoretical issues. And the students in UNM's Water Resources 571 class proved a patient and inquisitive audience as I pummeled them with lectures about Yuma and lettuce and my strange fascination with the fountains of Las Vegas.

Nora Reed insisted that I think carefully about who is marginalized in the decision-making processes. It was an important moral compass. William S. Reed isn't here to see the results, but without his loving support this book would not have been possible.

Emily Turner Davis at Island Press has been the best editor a writer could hope for, able to see through my mush to what I was trying to say, then helping me say it.

Most importantly, Lissa Heineman helped me realize this was a thing worth doing, and then listened to every thought and read every word.

CHAPTER 1

Rejoining the Sea

S TANDING IN THE DRY BED of the Colorado River at San Luis in the Mexican state of Sonora, just south of the Arizona border, Manuel Campa was insistent. The Mexican border city, perched on a low mesa to the east, is not just "San Luis." It is "San Luis *Río Colorado*." "It's the only city that has the name 'Río Colorado,'" Campa, technical director of the city's water utility, told me as we strolled the river's sandy bottom on a warm spring morning. It took imagination to grasp what Campa was getting at. Nineteenth-century steamboats once passed this spot. The Colorado was once a river here.

No more. The only thing capable of navigating the Río Colorado's bed that day was a four-wheeler with fat tires. Tamarisk, a scrappy invasive shrub, had long ago replaced native cottonwoods and willows along the river channel.[1] Water-loving beavers, once common, seemed a comic impossibility.[2]

Over the last century, we have taken the river's water, moving it through dams and canals to grow a hydraulic empire of farms and cities across the semi-arid Colorado River Basin. By the time the Colo-

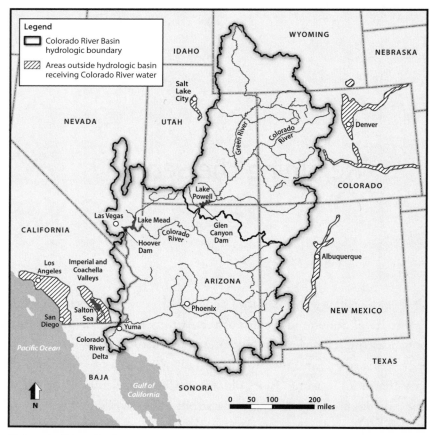

Colorado River Basin.

rado River approaches its feeble desert end, most of its water has been diverted to Denver, Albuquerque, Phoenix, Las Vegas, Los Angeles, Mexicali, and vast farmlands in between. Morelos Dam, twenty-two miles upstream from San Luis, diverts Mexico's share of the water—the last of the river—to the rich, productive farmland of the Mexicali Valley and cities to the west.

The first time I saw this, I was stunned. Driving the Yuma County levee past Morelos Dam in 2010, I saw the last trickles of water from leaks in the dam and a shallow water table disappear within a few miles into a sandy, dry channel. This great river, the Colorado, around which I have spent much of my life, whose water I have showered with and drunk, which has grown the food I eat and floated my boats for hundreds of miles, simply disappears into the desert sand.

But that spring day in 2014, Campa and I were awaiting something remarkable. In the midst of fourteen years of drought, with reservoirs dropping upstream and fears of water shortage gripping the Colorado River Basin, water managers were creating a modest "pulse flow," meant to mimic a natural spring flood through the desiccated delta.

It was a test of how much water would be needed for native plants to come back to life and repopulate the area. But as the water arrived at San Luis, and for the weeks after, it became something more. The usually dry riverbed past the town turned into a fiesta as children who had never seen water here frolicked in a briefly flowing Río Colorado. And at another, deeper level, it demonstrated unprecedented international cooperation to achieve a goal once thought impossible.

Chinatown

As California burned through the summer of 2015, its fourth year of drought, geographer Daniel Grant described the myriad photos of cracked reservoir mud and dried irrigation ditches and their accompanying headlines as part of a "genre of apocalyptic prophecy" that functions, according to Grant, "by diagnosing a human misalignment with nature, and foresees a future in which nature—as a kind of secular deity—punishes our errant behavior."[3]

For much of my professional life as a writer, chronicling our uneasy existence in this arid place, I embraced this narrative. When I wrote stories about drought for the *Albuquerque Journal*, we had an office joke about "the obligatory cracked mud photo." The paper's photographers and I would hover over stream-gauge data to find the driest stretch of

New Mexico's rivers, and it was always a bonus if they came back with an image that included a dead fish to punctuate the message.

These stories resonate, dominating our understanding of life in the arid West. Thus it is that the classic movie *Chinatown* has come to stand in for the history of Los Angeles water. It is a tale of villains bent on profit, messing with nature, and ultimately punished for their sins. To many, *Chinatown* represents water management in LA, despite historians' best efforts to remind us that it was just a movie, that things didn't really happen that way.[4] The apocalyptic vision isn't limited to fiction. James Lawrence Powell, in his 2011 Colorado River history *Dead Pool*, predicts a *Grapes of Wrath*–like exodus from Phoenix.[5]

Perhaps no work is more important to the West's narrative of crisis than journalist Marc Reisner's epic *Cadillac Desert*, ominously subtitled *The American West and Its Disappearing Water*. Its core message, one commentator wrote later, was that the overbuilding of dams and overuse of water in the western United States would "catalyze an apocalyptic collapse of western US society."[6] Whether that is a fair characterization of Reisner's work is an open question. *Cadillac Desert* has become shorthand for the water crisis, with all varieties of doom attributed to Reisner.

In fact, Reisner concentrated his fierce critique on what he saw as a corrupt process that overbuilt the West's great plumbing system. The subtitle notwithstanding, *Cadillac Desert* spends little time on the "disappearing water," or the actual human consequences of water shortages. But neither did Reisner shy away from apocalyptic rhetoric. In the 1993 afterword to the book's second edition, Reisner was explicit. California had just experienced what was at the time its worst drought on record, which, Reisner said, "qualifies best as punishment meted out to an impudent culture by an indignant God."[7]

Like many who manage, engineer, utilize, plan for, and write about western water today, I grew up with the expectation of catastrophe. I first wrote about water shortage in California during that same late-1980s–early-'90s drought Reisner bemoans. But as drought set in again across the Colorado River Basin in the first decade of the twenty-first century, I was forced to grapple with a contradiction: despite what Reisner had taught me, people's faucets were still running. Their farms were not drying up. No city was left abandoned.

I began asking the same question, again and again: when the water runs short, who actually runs out? What does that look like? Far from the punishment of an indignant God, I found instead a remarkable adaptability. In Doña Ana County on the Rio Grande in southern New Mexico, I saw farmers idle alfalfa and cotton fields, crops that bring low returns for each gallon of water, shifting scarce supplies to keep high-dollar pecan orchards healthy and productive. As water supplies dropped to record lows, farmers continued to prosper. New Mexico's cities fared just as well. In the midst of the drought, Albuquerque cut its per capita water use nearly in half, and the great aquifer beneath the city actually began rising as a result of a shift in supply and reduced demands. Across the Colorado River Basin, I found the same story over and over, from the fountains of Las Vegas and Phoenix to the farms of Imperial and Yuma counties, to the sprawling coastal metropolis of LA.

When people have less water, I realized, they use less water.

In spite of the doomsday scenarios, westerners were coping, getting along with their business in the face of less water. Things might have been easier had we not made the mistakes Reisner so ably documented, but we did what we did and, as scarcity sets in, we are adjusting to the new realities. I have witnessed this resilience time and again as I travel the hydraulic landscape of the western United States. This book chronicles my attempt to understand and explain where that ability to adapt comes from, how it works, and how we can call on it to get us through the hard times ahead.

The Myths

The catastrophe narrative isn't just inaccurate—it promotes myths that actually stand in the way of solving our problems. Most obvious is the myth that "water's for fighting over." The quote is wrongly attributed to Mark Twain, but it's also just plain wrong. Fighting rarely solves water problems, and scholars have found that collaborative agreements are far more common than winner-take-all fights, whether in the courts or with guns.

A corollary myth is that "water flows uphill toward money," as if the rich will inevitably run roughshod over their neighbors in the coming water wars. But a century of water allocation law and policy shows this is simply false for the vast majority of the water in the United States. One need simply look at farming's share of Colorado River water (some 70 to 80 percent) compared with its share of the economies of the US basin states (2 percent) to see the fallacy in that truism. The far-richer cities have far less water than their farming cousins.[8]

The most pervasive of the myths is that we are "about to run out of water." I've heard it countless times, and it usually follows a predictable pattern. Today, we need this much water to support this many people and this much farming. As either grows, we'll need more water, the narrative would suggest. When the "need" line crosses the supply line, we will "run out." This ignores history, where again and again we have seen both city and farm communities adapt and continue to grow and prosper without using more water, often, in fact, using less. But that deeply held fear of "running out" of water feeds back into the first myths, triggering a limbic response to protect "our share" against others.

And therein lies the risk. If everyone ignores their own adaptive capacity and simply fights for more, or even fights for the share they've got now in a shrinking system, we are led headlong into conflict, with dangerous results. If instead we recognize our ability to make do with less, and invest in institutions that facilitate water sharing, we can create systems for robust, flexible, and equitable water allocation. Only then can we preserve the West that we all have come to inhabit, know, and love.

Stories of Success

There are success stories in the recent history of the Colorado River's management.

The first type of success story involves communities that have found a way to use less water. It happens at many different scales and among many different types of water users.

Big cities, like Albuquerque and Las Vegas, have shown remarkable conservation success, with populations continuing to grow in recent decades even as water use goes down. Regional water managers at agencies like the Metropolitan Water District of Southern California and the Central Arizona Water Conservation District have developed innovative, flexible new approaches to managing supplies; storm-water capture, sewage reuse, aquifer storage, and other similar innovations have diversified sources of water and provided a buffer against drought.

Farm communities also have demonstrated the ability to do more with less, with agriculture thriving in California's Imperial Valley and Yuma, Arizona, even as they face pressure from outsiders charging that their water use is wasteful.

These stories are crucial in part because they show communities working with the cards they were dealt. When fighting over water, it is easy to say, "We can no longer afford to grow alfalfa in the desert," or "We can no longer afford a Phoenix or Las Vegas." Simply cutting off a few limbs would eliminate the deficit that causes the reservoirs to keep dropping. I reject that approach for two reasons. First, there are questions of justice and equity in deciding which communities stay and continue to use water and which communities must go. Second, though, and more important, we have no omniscient power giving us the ability to decide which water uses will continue. When we decide our future, the Imperial Valley and Las Vegas are at the table, defending their right to exist. As a result, the only tractable plans are ones that work with current water users.

That reality points toward the second type of success story. At the scale of the entire Colorado River Basin, in the messy complex of governance that manages decisions about who gets how much of the big river's water, we have made tremendous progress. Governments have banded together to come up with a plan to reduce California's overuse of the river, and they've developed a deal with Mexico to share surpluses and shortages, and even to spare some precious water to return flows to the Colorado's parched channel through its old Mexican delta. This was rarely easy, but it resulted in deals that benefited a spectrum of water users and community values.

Underlying both of these models of success is a willingness to recognize water problems and collaborate in solving them, often across geographic, political, and organizational boundaries. Conflict is sometimes a part of these processes, but ultimately success comes by avoiding fights over water.

But these successes have not been enough, something that can be seen most clearly in Lake Mead itself, the first great reservoir that stores the Colorado River's water for millions of people downstream. Despite the hope offered by the success stories described above, we have not done enough. Lake Mead continues to shrink. Water users continue to take more out of the Colorado River system than nature puts in, keeping us on an unsustainable path.

Just as importantly, westerners have failed to notice their own successes. Seeing Lake Mead drop should encourage water users to redouble their efforts, to build on the extraordinary conservation programs they have created. But too often, it instead triggers fear, fueling the myths in a dangerous feedback loop. If I think I need more water, I am more likely to be willing to fight for it.

Egret on the Lower Colorado River (© Lissa Heineman).

San Luis Río Colorado

Beyond the water banking agreements and lawn buyback programs and agriculture-to-urban transfers that will be needed to succeed is a fuzzy, at times difficult to grasp, often imperfect but nevertheless critical element to solving the Colorado River's problems. Participants have come to call it "the network."

The network is a sometimes formal but often informal group of lawyers, engineers, hydrologists, farmers, water managers, diplomats, and environmentalists who have been working together on these issues, often for decades. Meeting in conferences, on river trips, and in hotel bars, they must hold the seemingly contradictory goals of zealously guarding their own communities' water supplies while at the same time figuring out how we all can share as shortage looms.

This disparate group keeps what one participant has described as a "laser-like focus" on the big problems posed by the dropping reservoirs.

They are the ones who have to figure out how our society can live with less water. This is where our adaptive capacity must come from.

Social scientists who study big, complex, sprawling problems like this call such collective problem solving "network governance." The web of relationships among the people who carry it out is more expansively characterized as "social capital." It is the people themselves, but also their bonds. They have a shared understanding of the resources, of one another's needs, and of the complex set of rules that govern water's use. They have built trust and reciprocity over years of working together across difficult boundaries. Social capital is every bit as important and worthy of investment as physical capital: the pumps, dams, and ditches we build to manage and move our water.

The very existence of such a "network" can be problematic. What does it take to get admitted to the club, and who gets left out? What will that mean to the way we solve the river's problems? I recognize the risk that important stakeholders can be left out of the network, leaving their interests unprotected as the scarce water is divided up. But after years of studying management of the Colorado River I've come to believe that it is better than the alternative, which is the risk of constant conflict. A free-for-all could crash our system and leave someone the loser—it is hard to know who—without the water on which the community has come to depend.

The first time I wrote about Terry Fulp, a key manager with the Bureau of Reclamation, I described him as "the closest thing we have to a guy with his hand on the tap that controls the vast plumbing system built over the past century to distribute the Colorado's waters."[9] But I have come to realize in the years since I published that line in 2009 that, in reality, no one has their hand on the tap, and nobody has the ability to turn it down. Instead, we've built a decentralized system with no one in charge. This means that the only possible solutions are those that can emerge from the collaboration of the network.

That is why the scene I stumbled on one late afternoon in March 2014, in the sandy bed of the Colorado River near the Mexican town of San Luis Río Colorado, was so heartening.

For those brief few weeks, water was flowing and the residents converged on the normally dry riverbed for a rollicking, soccer-ball-kicking, four-wheel-driving, beer-drinking party. The reason was the network in action—a complex deal that opened the door to settling old water management controversies with Mexico, and won important benefits for US cities. And most importantly, it created a little pulse of water that returned the Colorado River to its historic channel past the city of San Luis. Remarkably, the Colorado River Basin's managers had all agreed to lower Lake Mead just a bit, to release some of their precious water, in order to bring a dead river channel in Mexico back to life.

I rolled up around suppertime after a day with a group of scientists and journalists. There seemed no finer place on Earth at that moment than the party beneath the San Luis Bridge.

There, amid the festivities, with no tie, barefoot, his suit jacket off and the legs of his dress pants rolled up, I found Terry Fulp with a group of friends, wading in the Colorado River. Fulp had been part of a stuffy official delegation earlier in the day, driven in armored State Department Chevrolet Suburbans to tour the newly flowing river. But at their last stop, Fulp bailed out of the official caravan to get a ride home with his friends, a group of environmentalists who had been working for more than a decade toward this day.

This was the network—old friends on opposite sides of what were once great divides: the conflict between environmentalists and the West's great water management agency, the conflict between Mexico and the United States, and the conflict between farms and cities for scarce water. They were standing, smiling themselves silly, in a river channel in Mexico, sharing a historic moment as a river slipped past them on a trip to rejoin the sea.

CHAPTER 2

Water Squandered on a Cow

THE WATER THAT IRRIGATES Corky Herkenhoff's alfalfa fields travels through dams and tunnels across two states to San Acacia, New Mexico. The path is long, but it is important to be clear about where it ends: as food for a cow.

San Acacia sits on the west bank of the Rio Grande. In other words, it already had a river of its own. But the natural flow of the Rio Grande is sparse and highly variable. So in keeping with the "you can never have too much" philosophy of water management in the West, New Mexicans used politics and money to grab more, building the San Juan–Chama Project to pipe water from Colorado River tributaries into the Rio Grande. Much of this imported water goes to the Santa Fe and Albuquerque metro areas, supplying showers, flushing toilets, watering golf courses and lawns. Some ends up in agricultural irrigation ditches. Some of that ends up on Herkenhoff's farm, growing alfalfa, most of which gets shipped out as feed to dairies on New Mexico's eastern plains.

As farming in the western United States goes, this is marginal stuff.

Alfalfa in the Palo Verde Valley, grown with Colorado River water (© John Fleck).

In this stretch of the Rio Grande Valley, the 2012 net farm income—the amount farmers made after deducting expenses—was negative.[1] Herkenhoff is relatively successful, but, explaining his family history at San Acacia, he joked that he is "the fourth generation to go broke on this operation." His father, who made a living as an engineer in the city, had advised against going into agriculture. "I didn't take his advice," Herkenhoff said.

Herkenhoff has tried all sorts of things on this land. One year he grew cantaloupes, but the market collapsed so he decided to feed them to his pigs. But the market for pork was lousy too: "That didn't work out for me, either." Over time, cheap land, cheap water, and a growing market in New Mexico's dairies has made alfalfa Herkenhoff's crop of choice.[2]

This is water subsidized, dammed, channeled, and diverted, so that it might be squandered on a cow.

On the grand scale of dams and diversions, the San Juan–Chama Project is tiny, taking less than 1 percent of the Colorado River's water.

But if you want to understand the problems facing the Colorado River Basin, you have to start at places like Herkenhoff's fields, and especially with alfalfa, because thousands of farms across the West are where most of the water goes.

The River

Rising from snowmelt on the spine of the Rocky Mountains, the Colorado River travels 1,450 miles, draining 246,000 square miles.[3] Physicists talk about "potential energy," embodied in the pull of gravity on an object. The higher something is—like snow, for example, resting in winter drifts high in the mountains—the greater the potential. For the snowpack in the Rockies, that potential is a gift of the sun. Lifted by the sun's energy from oceans into clouds, the water falls as snow on the high peaks, then melts and flows down, first by trickles, then joining into streams and rivers, turning the potential into actual energy of enormous power—to shape landscapes and, turned to human needs, to build a society.

It is easy to take this for granted today, but European settlers moving west across North America in the 1800s were encountering something new. Whereas in the eastern half of the continent the rain that fell from the sky was enough to build farms and towns, in the West the rain was sparse, the land was barren, and the water trapped in that snowmelt flowed through deep canyons. It would take the collective action of human communities to capture and channel the water, to put it to what law, policy, and history have come to call "beneficial use."

To do that required vast works of physical plumbing. The first major step was Hoover Dam, astride a deep canyon between Arizona and Nevada, completed in 1936. One of the great engineering achievements of its or any age, Hoover Dam was able to capture floods and store years' worth of water during wet times to supply downstream farms and cities during the dry times. Less than three decades later, the plumbers added

Glen Canyon Dam nearly 400 miles upstream, doubling the system's water-storage capacity.

Smaller dams and canals to distribute the water splayed out across the basin until, by the late twentieth century, we had turned a river into a grand water-distribution machine. Farm communities and cities, from Denver and Albuquerque in the east to Phoenix, Las Vegas, and Los Angeles in the south and west, grew up dependent on the water the Colorado River could now provide.

Alongside the physical plumbing, we built a vast institutional and legal apparatus to manage the water's distribution. In 1922, a "compact" among the states sketched out who was entitled to how much of the river's water. Legislation followed to clarify the water distribution rules, and more importantly, to authorize the massive federal subsidies needed to turn the western desert into something that resembled the cities and farms back east.

The Colorado River's problem is simple: there is not enough water to enable everyone to use the amount to which they are legally entitled. The river is like a company with contractual obligations to pay out more money than it is taking in. Sooner or later in such a situation there are two options: either a negotiated bankruptcy, in which creditors take less than their full entitlement and the company stays afloat, or the whole thing collapses. The former seems preferable to the latter, but negotiating such a deal is hard.

The roots of the problem lie in mistakes made in the early twentieth century. At the time the Colorado River Compact was negotiated, streamflow data was sketchy, based primarily on a single gauge at Yuma, Arizona, near the bottom end of the river system. But twenty years of data was sufficient to give the compact's negotiators confidence that they had at least 17 million acre-feet per year on average to work with, and likely more.[4] "The hydrographers and experts advise me that a twenty-year record on a river is adequate in its completeness and includes enough years to warrant an assumption that the average there deduced would be the average flow of the river in the future," Colorado lawyer Delphus Carpenter, the compact's chief architect, told his colleagues at the commission's meeting on November 12, 1922.[5]

That seemed like plenty. The commission allocated 7.5 million acre-feet of water for the states of the Upper Basin—Wyoming, Utah, Colorado, and New Mexico—and 7.5 million acre-feet for the states of the Lower Basin—Nevada, Arizona, and California. A treaty between the United States and Mexico signed in 1944 added another 1.5 million acre-feet of rights for water users in Sonora and Baja California, bringing the total to 16.5 million acre-feet.

What they did not grasp until years later was that their twenty-year baseline was unusual. How unusual? With the advent of tree-ring reconstructions of past climate, we now know it was the wettest twenty-year period in at least 500 years.[6]

As early as the late 1920s, before the first of the river's big dams was built, there were signs that the planners had made a mistake, that there was in fact less water in the river than they thought.[7] But political expediency trumped hydrologic doubt, and the institutional plumbers continued to rely on the old numbers as they divided up the river's flow and built the infrastructure to put the water to use.

For most of the next century, water use in the basin grew slowly enough to mask the problem. But as consumption grew into the original allocations and drought set in as the twenty-first century began, the shrinking reservoirs left the early institutional plumbers' mistakes on display for all to see. In the years since the compact was signed, the river has averaged just 15.5 million acre-feet per year, while water use has inexorably grown.[8]

The problems were compounded by a deep and persistent drought that began in the year 2000. Only three of the years from 2000 to 2015 had above-average streamflow, with overall flow during that period 16

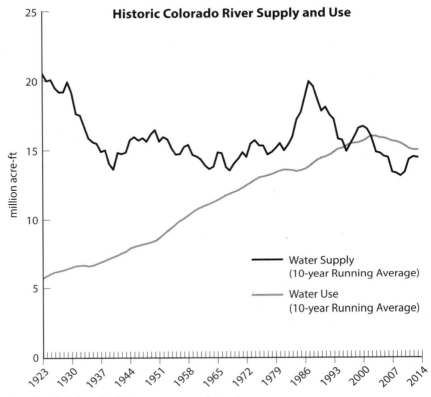

Graph of Colorado River water supply and use.

percent below the long-term average. Scientists first predicted in 1993 that climate change caused by rising greenhouse gases was likely to reduce the Colorado River's flow. As the studies projecting shortages mounted, by the 2000s it appeared evident that the change was under way. But despite the warning signs, the basin's big users continued to take their full allotments from Lake Mead. By the summer of 2015 the big reservoir had dropped to record low levels, with shoreline emerging that had not been above the waterline since the reservoir was first filled in the 1930s.[9]

The bank account was on the brink of being overdrawn.

Embracing Alfalfa

The most important legal principal undergirding the Colorado River's water allocations asserts that the first communities to put water to "beneficial use" get first dibs in times of scarcity. It is called "the doctrine of prior appropriation," and in the West, that most often means agriculture. The result is that half of the Colorado River's water is consumed by pasture and cattle feed crops, especially alfalfa.[10] Per dollar invested and gallon of water used, alfalfa in particular and pasture crops in general are the West's lowest-value major crop. But rather than seeing this as a problem, and arguing that it makes no sense, it's better to view alfalfa as part of the solution.

We can debate whether it was a good idea throughout the twentieth century to allocate so much water in this way. But that's done. Farmers like Herkenhoff in places like San Acacia made good-faith decisions about where to build their homes and how to make a living based on a national policy of subsidizing irrigation water and the infrastructure needed to deliver it. Change requires that we come to grips with the reality that the Colorado's history has made it a working, agricultural river, and that communities built their lives around those choices.

When the grand project that became the development of the Colorado River was launched in 1902 with the Newlands Reclamation Act, there was little notion that the water being impounded by the dams and distributed through the canals would fuel cities in the arid West. The plumbing was for farming, a concept embedded in the very name of the agency the law created, the US Reclamation Service. Money was "to be used in the examination and survey for and the construction and maintenance of irrigation works for the storage, diversion, and development of waters for the reclamation of arid and semiarid lands." *Reclamation* meant turning useless desert into productive farms. Francis Newlands, the Nevada congressman whose name the act bears, could not have imagined a Las Vegas metro area holding most of his state's population.

He thought irrigated agriculture was the only way to halt what was then a decline in Nevada's population.[11]

Thus it was that farmers, and the irrigation system managers who helped them, developed the basic architecture of the West's desert plumbing, the physical parts that moved the water from rivers as well as the institutional part that managed that process. Farmers told me this, often, when I came calling as a city dweller trying to understand their world. Long before the big inland cities were even a gleam in developers' eyes, farmers and their irrigation agencies, led by the Reclamation Service (later elevated to the "Bureau of Reclamation") were damming and diverting the water to make lives on the arid landscape. They built the superstructure on which our hydraulic empire was erected.

The best estimate is that between 70 and 80 percent of the basin's "developed" water—water turned out of the river for human use—goes to agriculture, much of that to alfalfa and other animal feed.[12] In the higher-elevation states of Colorado, Utah, Wyoming, and New Mexico, colder temperatures and shorter growing seasons make animal feed by far the most suitable crops to grow in all but a handful of farming valleys. In the Lower Basin states of Arizona and California (Nevada farm acreage irrigated with Colorado River Basin water is tiny), warmer temperatures and longer growing seasons expand farmers' options, but feed crops still take up a major portion of the irrigated land.[13]

An estimated 4.5 million acres of land in the US portion of the basin are irrigated at least in part with water from the Colorado River—an area the size of New Jersey. Total agricultural sales attributable to crops grown with Colorado River water were an estimated $5 billion in 2007.[14]

No crop meets the need for quality animal feed better than alfalfa, the "queen of forage." Its history as a cultivated crop goes back at least 6,000 years, to the region that is today Pakistan, Afghanistan, and Kashmir. Humans were almost certainly using it as a wild crop even earlier.[15] The history of alfalfa's earliest use in the United States is murky, but it took off during the California gold rush of 1849, when it became clear that, with sunshine and irrigation water, it could be an agricultural winner.[16] From the early years of irrigated agriculture in the Colorado River Basin, alfalfa has been an important crop. In 2014, Colorado River Basin farmers were growing alfalfa on 1.4 million acres of land, more than was devoted to any other crop.[17] Planted once, alfalfa can produce crops over multiple years. Cut and baled more than once a summer, it is stubbornly resilient in the face of drought and is a quality feed for horses, beef cattle, and dairy cows. Yet alfalfa and pasture grass provide the lowest agricultural return on water among the West's irrigated crops. In 2015, a University of California team estimated that vegetables and similar high-dollar crops return forty times more money per gallon of water than does alfalfa.[18]

And despite the low return, alfalfa uses a lot of water—some ten times as much Colorado River water as Las Vegas, by one estimate—because we eat a lot of burgers and pizza cheese, and the animals that produce them need something to eat.[19] That high use and low value has made alfalfa an attractive target in the struggle over allocation of the Colorado River Basin's water. City water users, scrambling to meet their needs in the face of scarce supplies and growing population, look at this and ask whether, in the twenty-first century, this allocation of water makes sense. "We all like farmers," Carl Boronkay, the director of Southern California's largest municipal water agency, said in 1991, "but there is no way people are going to be denied while farmers are spraying it on alfalfa."[20]

Unfortunately for the Carl Boronkays of the world, the solution is not as simple as pronouncing that the water should be given to rich city dwellers rather than squandered on cow food. Farmers like Corky Herkenhoff have good reasons for planting alfalfa. It is relatively easy and inexpensive to grow, and provides flexibility that other crops do not. As long as meat and dairy are staples of the American diet, the alfalfa to feed our animals will be part of our agricultural mix. It is often suggested that simply switching to vegetarian diets would solve our nation's water problems. But there is no policy lever to pull and make that happen. Most importantly, the allocation system set up a century ago gave farmers first dibs on the water—a system that has proven hard to change. We built checks and balances into our water allocation system that make it hard for someone like Boronkay to muscle in and simply grab the water away from farmers like Herkenhoff. The fact that alfalfa farmers hold more water rights than Las Vegas is proof that the old saw "water flows uphill to money" is wrong.

But if, rather than demonizing alfalfa, we embrace it, we'll find that the crop can actually help solve our water problems. A more expansive view of alfalfa and other low-dollar crops shows they play a key role in providing the adaptive capacity we need to adjust to a changing climate and changing values about how to use water.

Alfalfa makes our water system more resilient in three important ways. The first is the crop itself, which is wonderfully adapted to drought. It is a rich and easy-to-grow source of protein for animal feed, which is what makes it so popular. Its deep roots allow it to hunker down when water is scarce. That has long made it popular in places where water supply is variable. If the snowpack this year means you can't irrigate as much in the summer, that means fewer bales to sell, but not that your investment is lost.

The second advantage of alfalfa is its low value relative to other crops frequently grown in the Colorado River Basin. Compared to winter lettuce or durum wheat, you won't make as much money per acre of land and acre-foot of water with alfalfa. That means that when water becomes scarce, or there is a need to reallocate it, alfalfa is often the first crop to be fallowed, shifting water either to municipal users or to other, more valuable crops. It is a simple buffer if we offer farmers some benefit in return for relinquishing some of their irrigation water.

The third way in which alfalfa provides adaptive capacity is its portability. It is difficult under the laws governing the Colorado River Basin's water allocation to move water across state lines. But alfalfa is easy to move. When water runs short in New Mexico, a dairy can have its alfalfa shipped in from the Midwest. Bales of alfalfa on flatbed trailers provide one of the ultimate forms of flexibility in dealing with water shortages and allocation problems.

In order to capitalize on this flexibility, we need to develop institutions that both respect current water users and provide tools for moving water around more easily. Some of the changes happen on the farm itself, as farmers adjust to their shifting water supplies. But some require relationships—farm communities and municipal water managers building trust and developing new management rules that benefit both. Ultimately, that mutual understanding can lead to changes in the laws and institutions that allocate the basin's scarce water. There is no one big policy lever that can be pulled to make this happen, but there are many small switches that cumulatively will add up to the necessary change.

Farming the Desert

To understand alfalfa's complex role in helping to solve the Colorado River Basin's problems, we need to travel to the deserts of southern Arizona and California, where agriculture squeezes out the last drops of the river's US allocation before sending what remains to Mexico. Arizona's Yuma County and California's Imperial County, which flank the river here, began their modern agricultural life as alfalfa country. One observer in 1917 called alfalfa the "backbone of permanent fertility of Imperial Valley soil."[21] Today you can still find the crop here, but the region's transition toward more-lucrative lines of farming illustrates the opportunities, bringing more money to farmers and taking less water from the Colorado River.

Yuma is a historic Colorado River town. It grew on the southern side of one of the few good Colorado River crossings for a hundred miles in either direction. Miners heading to the gold fields of California in the mid-nineteenth century crossed there, and the Southern Pacific Railroad chose the bluffs that flank the river at Yuma for its rail crossing in 1877. But while life as a way station provided Yuma with an economic start, the real action looked like it would be in farming. The river flowed through a broad, flat valley that stretched fifty miles to the east, up the drainage of the Gila River. Yuma also had what seemed like an unlimited supply of water from the Colorado River slicing through the middle of the valley. But getting the plumbing right took decades. Early diversions left farmers' fields waterlogged, and flooding was an ever-present danger until the US government completed Hoover Dam.

Alfalfa was the principal crop grown in Yuma in the early years,[22] and by the 1940s nearly half of Yuma County's acreage was planted in alfalfa and other forage crops.[23] But over the last three decades, Yuma County's shift away from alfalfa has made agriculture there both more water-efficient and more economically productive. Alfalfa production fell from utilizing half of Yuma County's agricultural acreage in 1940 to one sixth in 2012.[24] In its place is a winter-vegetable empire that feeds the desires of the nation's consumers for year-round fresh salad greens. Yuma's farmers are making more money and using less water.

Across the river in California's Imperial County, getting the plumbing right was harder, eventually requiring a massive intervention by the US government to build an eighty-mile canal bigger than many western rivers to get water to desert farmland. By the late twentieth century, the majority of that water went to alfalfa fields.

Much of it still does. Imperial's turn from alfalfa has been less pronounced, but given its size as the largest single user on the Colorado River, it is no less important. In the early 2000s, 60,000 acres of alfalfa were taken out of production, with water shifted to other uses and some of the conserved water transferred to Southern California's cities.[25]

The shift away from alfalfa on the Lower Colorado has taken two forms. The first is simply the evolution of the farm economy. Yuma farmers in the last quarter of the twentieth century took up high-profit winter lettuce farming with enthusiasm. Water conservation was not the goal here, but rather a happy accident. The second is institutional. As regional water managers have come looking for opportunities to reallocate the Colorado River's scarce water, they have been able to make deals that take low-value alfalfa land out of production, moving the water to meet other needs and leaving the higher-value crops of the region's winter vegetable trade.

Isn't this the Carl Boronkay plan—to stop growing alfalfa and use the water for cities and more-profitable crops? The difference is in degree and agency. Farmers here are still growing some alfalfa. When there is sufficient water, they can grow more. When they grow less, it's by choice. Rather than being forced out of the market and facing dramatic economic upheaval, farmers can make a slower, more comfortable transition—often with sizable profits.

The Lettuce Empire

If you live in the United States or Canada, chances are good that the lettuce you eat during the winter came from one of these two counties at the tail end of the Colorado River. The sunny desert, good soil, and plentiful Colorado River water make the Yuma-Imperial farming district the perfect place—really the only place in United States—to grow lettuce in the dead of winter. Beginning in September, crews are sorting through orders from the big produce shippers, timing their plantings so that by November wave after wave of cauliflower, broccoli, iceberg lettuce, romaine lettuce, red leaf and green leaf lettuce, and celery are ready for picking. That precise schedule ensures that Kroger supermarkets and Taco Bell fast-food restaurants never go without. From December through February, more than 90 percent of the nation's lettuce comes from either Yuma or Imperial, with most of it passing through the refrigerated packing houses on a bluff above the Yuma County farm fields where it is grown.[26] As agriculture goes in the Colorado River Basin, this is some of the highest-value use of water, dollar per irrigated acre, that you will find.[27]

The farmers here and their irrigation district managers have turned the application of limited water onto rich desert soils into a combination of technology and high art. Tractors tow trapezoidal-wheeled metal devices known as "bolas" down irrigation furrows to tamp down the soil to exactly the right density and angle to efficiently move water between ridged rows of lettuce. Over the decades, they have tuned the length of irrigation run—the distance from the ditch turnout to the berm at the end of the field—to get the right amount of water to the plants without wasting any. Where runs three decades ago might have been as long as half a mile, you will rarely see one now longer than 600 feet.[28]

The whole enterprise represents a complex system of skilled labor, deft irrigation, and integrated production and distribution that has turned these valleys over the last forty years into some of the highest-value agricultural land in the country. Water use has gone down because the high-intensity irrigation has shifted to the cooler winter months, while agricultural revenue has gone up.[29]

In the mid-1970s, at the peak of Yuma County's water use, farmers there consumed more than 967,000 acre-feet of water, nearly as much water as the entire share diverted west to Los Angeles. Half of the valley's irrigated farmland—114,000 acres—was planted in alfalfa or cotton, also a crop that returns little money for each gallon of water consumed. By the early 2010s, the acreage devoted to alfalfa and cotton was cut nearly in half and the amount of water consumed dropped by nearly a third.[30]

In that time, total agricultural sales in Yuma County rose from $900 million in the mid-1970s (in inflation-adjusted dollars) to an average of

$1.2 billion per year since 2010.[31] Yuma County (and, as we shall see, other communities like it) are evidence that flexibility in the agricultural system provides a big part of the adaptive capacity we need to make our way through a water-scarce twenty-first century.

Deficit Irrigation

Often a straight transition away from alfalfa is the best choice for farmers and the basin as a whole. But in other instances, the cutbacks can be more complex. Just up the river from Yuma, in the Palo Verde Valley, we see how alfalfa may fit into the region's future. Palo Verde has some of the oldest farmland on the river, with some of the most valuable senior water rights. In 2007, it played host to a promising water-conservation experiment. Working with local farmers, University of California researcher Khaled Bali tried simply not watering alfalfa fields during the heat of summer.

Palo Verde has long been an innovator. As one of the first places in the lower desert where European settlers tried farming in the Colorado River Valley, it has some of the highest-priority water rights on the river, meaning it will be one of the last places to go dry in drought. That has made it an extremely attractive place for urban water agencies looking for deals, which is what makes Bali's experiments so promising.

This is some of the hottest country in the United States, with average daily temperatures from June through September above 100° F (38° C). In weather like that, it takes a lot of water to grow alfalfa. Importantly for Bali's purposes, it is so hot that the summer yield in alfalfa crop goes down, even as water use is going up. But recall the remarkable properties of alfalfa. Its deep roots allow it to hunker down in summer. If you deprive it of water, your yield will drop even more. But the crop does not die.

What Bali found when he experimented with side-by-side fields, some irrigated during summer and some not, was that yield went down and water was saved. This is an unsurprising result. Farmers have known for years that when their water supply runs short, they can get away with skipping an irrigation cycle and their plants will survive. They'll just have fewer bales to feed to their cattle or send off to their dairy industry customers.

Bali's experiment extended that to the idea of *intentionally* cutting short the water supply rather than simply doing it out of necessity. This opens up interesting policy options. If, say, the water saved was sufficiently valuable to a municipal water user to compensate the farmer for the lost yield in return for sending the saved water off to the city, you'd have the room for a deal that keeps the land in production and the farmer in business while also providing water to the city.

This sidesteps the biggest criticism of agriculture-to-urban water transfers—that they dry up land and community livelihoods. "Buy-and-dry" has become an epithet in the Colorado River Basin, and deficit irrigation provides an alternative if we can get the institutional arrangements right. By one calculation, widespread use of intentional deficit irrigation in alfalfa fields in Arizona and Colorado irrigated with Colorado River water could save nearly four times as much water per year as the annual consumption of the Las Vegas metro area.[32]

If this is so straightforward, why do we not see it happening? When I talked to Bali about his alfalfa deficit-irrigation experiment, he made a crucial point: "The science is the easy part."[33] As we will see, getting the institutional infrastructure right—arranging a deal between willing buyers and sellers, agreeing on a way to measure the saved water and get it to the alternative uses, changing the rules so it can be moved from one place to another—is a much harder problem.

One Alfalfa Field at a Time

Corky Herkenhoff would be happy to work out a deal, for a price, to share some of the Colorado River water he currently spreads on his alfalfa fields with one of the central New Mexico cities to his north.[34]

So far, the cities have not been parched enough to need to work out such a deal, but it is easy to imagine what it might look like: In a dry year, the city pays Herkenhoff to lay off the irrigation, sending some of his water to the city's municipal treatment plant. Herkenhoff takes the money and heads to Florida to go fishing. The barriers here are in the rules and procedures—the institutions through which we manage the water. Currently, New Mexico water law and policy don't allow a deal like that. But as water gets scarcer, pressure to reduce those barriers will grow.

How much US agriculture might miss a reduction in Colorado River Basin alfalfa remains an open question. One of our most flexible crops, it is currently grown in forty-two of the fifty US states, with places like Wisconsin and the Dakotas growing far more of it than the farm communities being irrigated with Colorado River water.[35] Hay has become a global commodity, one that is readily shipped long distances both within the United States and overseas. Dairies, the largest consumers, showed remarkable flexibility in shifting to other feeds as California's alfalfa production dropped during the drought of 2012–15; milk production stayed relatively stable in the state.[36] And using rail transport, growers currently ship alfalfa all the way from Idaho to Florida and Kentucky.[37] That diversity suggests plenty of adaptive capacity in the US farm sector to provide the feed crops to meet consumer demand for burgers and pizza cheese, though there remains a risk that the notoriously footloose dairy industry could in the long run migrate to be closer to reliable sources of feed.

This and the examples provided above show that while agriculture currently uses the bulk of the Colorado River Basin's water, it is possible to scale that back in a way that conserves water to respond to shortfalls while preserving the integrity of the communities that grew up around our old ways of allocating and using the river's flow.

But with no one in charge of the whole system, and therefore able to impose solutions, they have to be implemented painstakingly, one farm district and municipal water agency at a time. That is the project ahead of us.

CHAPTER 3

Fountains in the Desert

"You ALMOST HAVE TO HAVE grown up in the Old West to understand," explained a Las Vegas water manager to a visiting East Coast journalist in the early 1990s. "There is an emotional, almost irrational attachment people have to water out here. Their whole existence is tied to water."[1]

So it is that the driest big city in the United States cultivates an image of being the wettest. When R. E. Griffith built the splashy new Last Frontier hotel and casino in Las Vegas in the early 1940s, he put the swimming pool in front.[2] He wanted it to be the first thing motorists saw as they streamed into town on US 91, finishing the hot drive from Los Angeles to what was then an artificial oasis in its infancy.

Down the road from where the Last Frontier once stood, on a warm desert evening seventy-five years later, I joined a crowd along South Las Vegas Boulevard to watch the famed fountain in front of the Bellagio Hotel. I could see why Griffith's insight has endured. Like the Last Frontier, the Bellagio features its fountains in front, with a wide quasi-European boulevard sidewalk to add to the dramatic effect. All evening long, crowds gathered, mesmerized by the fountains' musical

The fountains at the Bellagio Hotel, Las Vegas (© John Fleck).

extravaganzas. The glittering gambling palaces of the Las Vegas Strip are full of similar displays—vast tiled swimming pools at Caesar's Palace, the Polynesian wonderland of the Mirage, the pirate ships of Treasure Island, the fake canals of the Venetian. But there is no greater icon of Las Vegas and its uncomfortable relationship with water than the fountains in front of the Bellagio Hotel.

Nor is there an icon more misunderstood.

In a desert that averages four inches of rain per year, the fountains' pond will never be mistaken for Lago di Como, the northern Italian lake on which the luxury hotel's namesake sits. The original Bellagio enjoys a balmy Mediterranean climate, averaging forty-two inches of rain per year. In hot, dry Las Vegas, weather has never been the chief draw.

But the Bellagio's fountain, often mocked as a symbol of water excess in the arid Southwest, may in fact represent some of the highest-value water around. The 12 million gallons a year needed to keep it topped up starts as water too salty to drink, drawn from an old well that once irrigated the Dunes Hotel golf course. Twelve million gallons sounds like a lot, but it's really just enough to irrigate eight acres of alfalfa in the Imperial Valley.[3] Total revenue at the seven giant casino–resort hotels contiguous to the fountain, at the corner of Flamingo Road and South Las Vegas Boulevard—the heart of the famed Las Vegas Strip—is an estimated $3.6 billion.[4] Include all of the hotel/casino operations in the greater Las Vegas metro area, and the total rises to $21 billion.[5] That compares with total agricultural revenue of $1.9 billion in all of Imperial County.[6] Imperial County's farmers get ten times the water Las Vegas gets. Las Vegas makes ten times the money Imperial County farming does. Given the crowds lining the sidewalks for each one of the fountain's dancing-water shows, the fountains must represent one of the most economically productive uses of water you'll find in the West.

This is the paradox of Las Vegas—that the city with the least water to work with, that has long been closest to the edge of the water-supply cliff, is the most ostentatious in the display of water. This has made Las Vegas an attractive target, because it is the focus of an almost pathological loathing on the part of some who view it as the great emblem of the West's sin against nature, and who view water scarcity as the tool by which that sin will be punished. Urban critic Mike Davis exemplified the form when he called Las Vegas "diabolical" and suggested what he called "the Glitterdome" was headed for an "eschatological crackup."[7] Las Vegas may have turned into the sort of urban agglomeration that critics like Davis find sinful, but a close look at the math behind its water policies, supply, and usage suggests that Las Vegas water managers have done a credible job of staving off their sin's punishment. Far from a path to destruction, they have developed the needed tools, and they have the track record demonstrating the use of those tools, to allow Las Vegas to follow the path the community's leaders have chosen.

To understand how to solve the Colorado River Basin's water problems, we have to come to grips with the illusion and the reality of Las Vegas water.

Fountains in the Desert

In terms of water, Las Vegas is handicapped on three counts. First, measured by rain falling from the sky, it is the Colorado River Basin's driest city. Second, an accident of history left it with the smallest allocation of Colorado River water to supplement that which nature fails to provide. Third, its reputation for hedonism blinds us to a reasonable assessment of what it actually is and does.

Combine its stark aridity with an economic model based on the illusion of lushness and a seemingly insatiable desire for growth, and you have what looks like a recipe for disaster. But Las Vegas will fool you. "We do fake well," one of the professional foolers, a landscaper who specializes in the town's art of deceptive lushness, told me as we cruised its boulevards looking at medians and apartment complex gardens.

That veneer of water—the fountains, palms, and ponds of the imitation tropical/Mediterranean/Caribbean Strip—is an illusion, promising much but doing it with very little water. In fact, a successful water conservation effort and its underlying governance structures have made Las Vegas a model of progressive water management. To a greater extent than any other city in the West, Las Vegas decided in a very public way what it wanted to be, how much water this would take, then laid out a plan to make this happen. Las Vegas's leaders chose growth, and the Southern Nevada Water Authority has made it possible.

Despite the rhetoric of imminent doom, the math is inescapable. From 2002 to 2013, the greater Las Vegas metro area grew by 34 percent to a population of more than 2 million people. During that same period, its use of Colorado River water—its primary source of supply—dropped by 26 percent. By the second decade of the twenty-

first century, Las Vegas had become a leading example of a phenomenon that has changed water management across the United States—
the decoupling of water use and growth. As population, agricultural
productivity, and economic activity in general keep rising across the
United States, water use does not.[8] By the mid-2010s, Las Vegas was
booming, yet the metro area was *not* taking its full supply from the
beleaguered Colorado River.

It did this in part through an aggressive conservation effort, which
reduced per capita water consumption by 40 percent. That impressive
number doesn't change the fact that Vegas remains a profligate user of
water compared to other western cities. But it does suggest an opportunity. Changing deeply embedded community behaviors and attitudes
toward water takes time, and Las Vegas got a late start. The fact that it
still has a long way to go means that, if Las Vegas wants to keep growing
and is willing to make more changes, there remains room for its population to grow within its current supplies.

This was not preordained. In the 1980s, the municipal water agencies
scattered across the Las Vegas Valley were involved in a classic "tragedy
of the commons." The state had not used its full allocation of Colorado
River water. Each utility was racing to use more before Nevada hit its
limit as the metropolitan area grew.

Las Vegas leaders were up front about the challenges they faced. If
they wanted to grow into a super-city, they needed to figure out how to
do it within the constraints of a limited water supply. Neighbors would
have to stop competing and figure out how to share. The first step was to
band together to create, in 1991, the Southern Nevada Water Authority,
a unified super-agency to oversee the distribution of their sparse allocation of Colorado River water.

The second step was an aggressive but voluntary conservation program. The Water Authority paid homeowners to tear out old lawns, and
it placed tight controls on landscaping in new construction. In doing

so, Las Vegas was making peace with its own environmental psychology. In the stark desert, neighborhood greenery provides a sense of comfort and safety as the city stares at the bared teeth of a hostile world. Las Vegas did not abandon that motivation. It simply decided to pursue it with less water. Casino fountains were not forbidden. Instead, they were required to switch to brackish groundwater rather than spraying precious Colorado River water into the hot night air. The Valley's political leaders pushed the conservation message hard with a public relations campaign that changed attitudes. And Las Vegas capitalized on its proximity to Lake Mead, treating sewage and returning the effluent to the reservoir so that it could be fully reused.

These measures worked. Las Vegas water consumption began a significant decline, even as its population continued to rise.

While there is much to be learned from the specific steps that Las Vegas took, one of the biggest lessons lies beyond things like lawn removal programs and effluent reuse. Those sorts of specifics will vary from region to region. But the underlying principle applies across the board—it takes governance and changing community attitudes to make things happen.

The Springs

Even while they flout nature with imported water, every city in the arid southwestern United States is anchored to its first water. Los Angeles grew around the Los Angeles River, where farmers could first spread its water across the arid coastal plain to water their crops. In Denver, it was the South Platte. Albuquerque grew up on a valley floor where the Rio Grande emerged from narrow canyons to the north into land with a combination of water and a climate warm enough to grow food. You can trace local history by starting with a community's first supply of water.

But while Las Vegas's most famous water is the nearby Colorado

River, the river's water was at the bottom of a deep canyon, inhospitable and largely inaccessible. To find Las Vegas's first water, you instead have to drive four miles north of the fountained casinos of the Las Vegas strip to a community park flanked by a suburban neighborhood and a shopping mall. There, an old *cienega* (spring) now called Las Vegas Springs Preserve marks one of the many places in the Great American Desert where water seeped to the surface on its own. Groundwater, accumulated over millennia in the sands and gravels of an aquifer beneath the valley floor, found a path to the surface at the cienega. Such aquifers are a common and important feature of the valleys where the West's cities emerged, and oases like this served as geographic organizing principles for the first wave of humans to make a living on the hot, dry landscape. If you wanted water in the desert valley, you had to come here.

For at least a thousand years before European immigrants arrived and began changing the landscape, Las Vegas Paiutes used the cienega, with its reliable supply of water, as a winter home. Historians think it likely that early Spanish travelers found their way through the valley in the 1700s, but the first sketchy written record does not show up until Rafael Rivera, a scout in the employ of New Mexican trader Antonio Armijo, wandered up Las Vegas Wash out of what is now Lake Mead and into the valley that would become Las Vegas.[9]

Water drew trading parties, hopping from one desert spring to the next, but beyond those temporary visitors and a brief and quickly aborted Mormon settlement, the Paiutes largely had the place to themselves until 1861. That year, the Comstock mining boom brought the first wave of European immigrant settlement that pushed the Indians out once and for all.

The railroad arrived in the early 1900s, remaking Las Vegas just as it did every place it reached in its march across North America. Seizing the opportunity, one of Las Vegas's earliest real estate entrepreneurs, J. T.

McWilliams, bought up land on the west side of the tracks and began selling lots. McWilliams seems to be first in a long line of Las Vegas Valley boosters who saw an opportunity in the burgeoning population of Southern California and ran ads in Los Angeles newspapers to lure Angelenos with a "plentiful supply of the purest water." One had to sink a well but twelve to twenty feet, McWilliams told his potential customers.[10]

The Union Pacific's land development subsidiary, the Las Vegas Land and Water Company, delivered most of the water to the early town through a network of cheap, leaky redwood pipes, but the leaders of the growing community needed more control over the supply. Thus Las Vegas's first experiment in water governance emerged: the Vegas Artesian Water Syndicate. It repeated a pattern seen frequently across the West—the need for collective action to overcome aridity.

The syndicate's planners misunderstood why they needed water. Like most who moved to the arid western half of the continent during this time, they thought the future lay in farming. "We have many thousands of acres rich enough for farming, and level enough for irrigation," they wrote in their 1905 prospectus inviting investors. "We believe that artesian water may be had in abundance."[11]

Farms still dot the narrow river valleys of the Muddy and the Virgin Rivers an hour's drive east of Las Vegas, but agriculture never took off in the Las Vegas Valley in the same way that it did in other parts of the growing West. Still, while the early developers' agricultural projections were wrong, they were right about the need for, and value of, water in the desert. The artesian wells that the syndicate promised were critical to the early development of the valley.

Hoover Dam

That is largely where Las Vegas stood until the 1920s—a small rail stop and desert outpost with little to grow an economy and therefore no need for water beyond that which the valley's limited aquifer could provide.

The accident of Las Vegas's geography, just miles away from the deep canyons of the Colorado River, was about to change that. The untouchable water was within reach, but the Las Vegas of the 1920s could not begin to grasp its implications.

"Action of 7 States Means Millions to Las Vegas," the *Las Vegas Age* proclaimed on November 25, 1922, as it formally announced completion of the Colorado River Compact. The millions would come from building a dam that, thanks to "the Hand of Destiny," would surely be built at the ideal dam sites in the canyons southeast of town. The *Age* also trumpeted the importance of cheap power, which would help Las Vegas compete with big industrializing cities back East. If any thought was being given to the water supply a new dam might provide, the newspapers of the day did not mention it.[12]

Watching Las Vegas's twenty-first-century water struggles, it is hard to imagine a time when there was plenty. But throughout most of the first half of the twentieth century, the springs and the aquifer that fed it provided an ample supply.[13] Until after World War II, there was little connection between Las Vegas and the Colorado River. When Arizona, California, and Nevada negotiated the 1928 Boulder Canyon Project Act, it was jobs from dam construction, not water from the river, that Nevada wanted. With almost no agriculture, little industrial base, and no inkling of the resort mecca that would blossom in the decades to come, Nevada settled for 300,000 acre-feet of Colorado River water, 4 percent of the total available to the three states and just one-fifteenth of powerful California's share.[14] "Nevada consistently took the position, accepted by the other States throughout the debates, that her conceivable needs would not exceed 300,000 acre-feet," wrote Justice Hugo Black years later in a US Supreme Court Colorado River water-allocation decision.[15]

Las Vegas benefited from bursts in federal spending in the 1930s as the nearby Hoover Dam was being built, during World War II, and in the early Cold War years that followed. A safe distance from the coast and the imagined risk of Japanese bombers, Las Vegas was home to Basic Magnesium, a war factory that provided material needed for airplane bodies. The plant put down roots for what would become the working-class suburb of Henderson. Basic Magnesium also provided the initial plumbing to bring Colorado River water to the valley.

By the 1940s, as World War II brought a second wave of spending for a military base and war factory, and a wave of new tourism flowed from Los Angeles, Las Vegas ran up against the same limits that had hit other western cities before it: Las Vegas had burned through its groundwater in a hurry, and the aquifer on which it had depended was no longer sufficient to meet its needs.[16]

Historian Eugene Moehring describes a 1944 meeting at which members of the Las Vegas chamber of commerce, gathered at the El Rancho Vegas, sat "in stunned silence" as state engineer Alfred Merritt Smith warned that groundwater pumping was putting such a strain on the aquifer that he might block some new well drilling, a move Moerhing said "would have crippled Las Vegas's postwar growth." The only solution, Smith said, was to pipe water up from Lake Mead to serve the rapidly growing metro area's needs.[17] Previously, Las Vegas had stood apart from the great network of Colorado River water users that already stretched from Denver to Los Angeles. That changed on the day Smith warned Valley residents that they were running out of water.

The first leg of a pipeline to bring Colorado River water from Lake Mead, over the hill and into the valley, had already been built in the 1940s for Basic Magnesium. Canny civic leaders, already heeding inklings that Las Vegas might need Colorado River water, had persuaded the plant's builders in 1941 to build a water line larger than was needed for the factory itself.[18] But the scale was nevertheless limited, and Smith's warning suggested that a much larger system would be needed.

At that time, the Valley's water needs were modest enough that the old Las Vegas Land and Water Company was sufficient to get the job done. But by the 1940s, the metro area's political leaders realized that they needed to build a new piece of institutional plumbing, a government agency responsible to the community as a whole rather than an appendage of the private sector. Once again, the early water-governance structures proved inadequate to meet the water needs of a community making the jump from rural agriculture to modern metropolis.

The newly formed Las Vegas Valley Water District bought out the railroad's interests and began laying pipe to connect Las Vegas proper with the old Basic Magnesium system in Henderson. Colorado River water finally began flowing to Las Vegas.

Learning to Work Together

When (formerly) sane water managers open the fire hydrants and run water down the streets of a desert city, you have a problem. By the 1980s, southern Nevada had a very big problem. Federal help brought Colorado River water over the hill and into the valley for the growing city, but figuring out how to put it to good use turned out to be harder than anyone expected.

In 1965, Congress passed legislation authorizing construction of the Southern Nevada Water System, an expansion of the old Basic Magnesium pipeline that would finally give the Las Vegas Valley the infrastructure needed to exploit its 300,000 acre-foot share of Colorado River water allocated under the interstate water deals made in the 1920s. Compared with the amount of water being pumped to Los Angeles, this remained a tiny slice of the Colorado River pie, but it finally seemed like enough water to secure Las Vegas's foothold in the desert.

Beyond the 30,000 residents the valley's water supply had supported in the 1940s, Las Vegas leaders could now see a path to the creation of a major city in the desert. Six pumping plants would move water uphill from Lake Mead through a four-mile tunnel and more than thirty miles of underground pipes. Las Vegas was finally joining the greater Colorado River Basin plumbing system.[19]

Nevertheless, over time Las Vegas quickly found that, while it had the physical plumbing it needed, the metropolitan still area lacked the institutional plumbing to properly allocate and manage the now-substantial shared water supply. The formation of the Water District had been a start, but it wasn't enough. Instead of sharing the water, the communities of the Valley engaged in what historian Christian Harrison called "hoarding and wasting" as each raced to use more water and thereby lock in a larger share of the scarce supply.[20]

The problem lay in the rules used to decide who got how much. The state's Colorado River Commission determined the formula, which was based on each municipality's previous year's usage. That created incentives to maximize water use rather than conserve it. What we call "Las Vegas" is really a hodgepodge of smaller municipalities and unincorporated areas, which at the time was served by a hodgepodge of water purveyors. Four cities within the Las Vegas metro area—Las Vegas itself, North Las Vegas, Henderson, and Boulder City—as well as the unincorporated areas of Clark County were all racing to use more water to support more development, which would allow them to grab the largest-possible piece of the expanding tax base.[21] In one famous case, Boulder City (the affluent and casino-free enclave closest to Hoover Dam) actually opened fire hydrants and dumped water in the streets in order to ensure a large allocation the following year.[22] Nevada was still not using its full 300,000 acre-foot share of Colorado River water, which ought to have been a good thing. But rather than having a rational discussion about the best way to accommodate rising water needs as the region grew into the allocation, individual communities were scrambling for their piece of the pie.

Harrison ties Las Vegas's moment of transition to the 1989 opening of the aptly named Mirage, the first of the multi-thousand-room mega-casino-resort giants. It launched a highly profitable building spree, and southern Nevada's use of Colorado River water began to shoot up, doubling between 1985 and 1993.[23]

Beginning in 1990, the Valley's water providers took the tentative first steps toward collective water management. As we will see later, a shared understanding of resources is necessary for cooperation, and nowhere demonstrates the point better than Las Vegas. The water providers hired a consulting firm, Water Resources Management, to provide a credible, independent answer to the question of how much water Las Vegas had and how much it needed. The answer, in WRMI's 1991 report, was scary. On its current trajectory, Las Vegas growth would overshoot the Valley's water supply within five years.[24]

While discussions were under way to deal with the problem, the Las Vegas Valley Water Authority (one of the key purveyors, which delivered water in and around the city of Las Vegas itself) took a step unheard of in growth-oriented Las Vegas by issuing what amounted to a unilateral halt to the extension of water service to new developments. The Authority sent a clear message to the development community: if you want to keep building, Las Vegas needs to get its water house in order. It also sent a message to the other Valley water agencies that the days of racing to lock up water-use allocation and development-driven tax revenue were past.[25]

By July 1991, the Valley's seven largest water agencies had come to an agreement on the formation of the Southern Nevada Water Authority. The new agency was an umbrella institution that acted as a middleman between the Bureau of Reclamation and the local water agencies to distribute Colorado River supplies. Rather than fighting over allocations, they agreed to pool their water rights. When there were shortages, the pain would be shared.[26]

Make the Desert Blossom Like a Rose

Policy machinations matter, but it's ultimately people who will determine Las Vegas's water future. You can see it in the conservation ethic of landscaper Kurtis Hyde. Hyde calls himself a "gardener," but that description hardly captures his role, or the importance of people like him, in the West's evolving relationship with water.

Hyde, the son of a schoolteacher, grew up in a little northern Nevada town called Lovelock, on the interstate between Reno and Salt Lake City. He was born into the Mormon faith, which is an important piece of his story. Since they arrived in Utah in the late 1840s, Mormons have been central to the task of measuring, marshaling, and using the West's water in new ways. From childhood, Hyde embraced Brigham Young's edict, based in the Bible's book of Isaiah, to "make the desert blossom like a rose." "I've been a gardener since I was a little boy," Hyde told me as he sat behind a desk at Par 3 Landscaping, a company that sits at the intersection of water and Las Vegas's future.

A graduate of Brigham Young University with a degree in botany, Hyde thought he was headed to a career in the Forest Service, but he ended up instead at the forefront of Las Vegas's water-conservation industry. Water bills in Las Vegas remain modest, but for businesses they can add up, which creates an economic incentive to conserve. Perhaps more important, though, is public awareness of the city's fragile water supply, which in the past few decades has created an increasing conservation ethic. Because most of Las Vegas's indoor water use is fully recycled, running out of the valley's sewage treatment plants and back into Lake Mead where it can be reused ("We're only borrowing it, and we have to give it back," one Las Vegas executive explained), most of the region's conservation efforts have focused on outdoor landscaping. There, water pumped up from Lake Mead evaporates or is transpired from the leaves of plants, leaving pleasant greenery in its wake but contributing to the growing uneasiness in this desert city.

Hyde thinks it's fine—even necessary—for the resorts to be lush. They are the core of his adopted city's economy. It's out in the spreading suburbs where the water conservation action is.

Driving west through town, Hyde points at median strips and shopping center borders that used to be lawn. Green grass has its place, Hyde said as he pointed to a little grassy play area in a West Las Vegas neighborhood that his company helps maintain. You want someplace for the kids to throw a Frisbee or kick a ball. But you shouldn't waste precious water on a strip of grass that no one can use.

Par 3 replaces sprinklers that used to spray water into the air, evaporating a significant portion, with drip systems, some that detect moisture need with weather sensors that adjust watering schedules to meet optimum conditions, maximizing every drop of water consumed. So Hyde is still pursuing Brigham Young's edict. He just realizes that it has to be done with less water.

And the conservation Hyde is practicing has been central to the Las Vegas strategy. After the formation of the Southern Nevada Water Authority in the early 1990s, there was a real effort to slow growth of the Valley's water use, but it was not enough. By the early 2000s, Lake Mead was dropping and Las Vegas had finally hit the limit of its Colorado River water allocation.

Recycling sewage, an option considered by many western cities, wouldn't help, because Las Vegas had long ago maxed out its use of that tool. Treated sewage has long flowed down Las Vegas Wash and back into Lake Mead, where it can be used again. In coastal cities, putting your sewage effluent to use rather than dumping it in the ocean is an important water-policy tool, but Las Vegas was already using it. That also meant that while indoor plumbing—low-flow shower heads and the like—got some attention, most of the conservation action was outdoors. A gallon saved in the toilet meant a gallon less pumped out of Lake Mead, but also a gallon less returned. There were some advantages because of savings in water treatment costs and energy, but the overall impact of indoor conservation was close to net zero on the overall Las Vegas water supply.

Outdoors, though, water put on lawns and gardens is lost to the system forever—"consumptive use" in the jargon of the water manager. So the Water Authority paid homes and businesses to tear out lawns—4,000 acres' worth since the program was launched in the early 2000s. This period was the depth of a frightening drought. In a span of four years, nearby Lake Mead dropped eighty-four feet. Drought was tangible to the residents, and they reacted by tearing out lawns and taking other water-saving steps with a surprising vigor. Per capita water use dropped 20 percent, and then kept going. Las Vegas's use of Colorado River water dropped 26 percent by 2013, even as its population went up by 34 percent.

Residents are restricted to outdoor watering just once a week in the winter, and three days per week in the summer, and if you water on the wrong day, you'll hear from your neighbors.[27] In new homes, front yard lawns are prohibited entirely, and backyard lawns are limited to 50 percent of landscaped areas. In nonresidential development, lawns are generally prohibited entirely. When I walked by the famous Mirage, the resort casino that started it all back in 1989, I noticed that the lawn out in front is made of fake grass.

A Multi-Purpose Agency

As often happens with the creation of such collective water-management agencies to act on behalf of a region as a whole, the Southern Nevada Water Authority became more than just a water wholesaler. Following a pattern seen many times across the region—with the Metropolitan Water District of Southern California and, in particular, the Central Arizona Water Conservation District—the SNWA became the umbrella agency for the region's water management as a whole. Issues that were beyond the scope of any one community—regional conservation, sewage reuse, expansion of the Southern Nevada Lake Mead intake system, protection of the aquifer, acquisition of additional sources of water from other parts of the state, and especially negotiations with the federal government and the other Colorado River Basin states—could be taken up by the SNWA, making them Las Vegas–wide problems and thus expanding the range of policy options available to solve them.

In addition to water conservation within Las Vegas, the agency was aggressive in pursuing new water supplies. Most controversially, it filed water-rights claims to what it called unused groundwater across vast swaths of rural Nevada. The idea was to pump the groundwater and pipe it to Las Vegas. This triggered a legal struggle, with rural residents and Nevada's neighbors in Utah fighting what critics view as an old-school urban grab of rural water reminiscent of Los Angeles buying up groundwater from the Owens Valley of rural eastern California in the early twentieth century. They disputed the Las Vegas claim that the water was "unused," saying the groundwater pumping would dry up natural springs and leave desert in its wake. This was the sort of project that would have been impossible for any of the formerly squabbling Las Vegas metro-area water agencies to pursue on their own.

The project, while still alive on paper, was largely stalled by the early 2010s, in part by legal challenges, in part because of the enormous costs, and in part because slower population growth and Las Vegas's remarkable water-conservation success made the rural groundwater unneeded for the time being. By the summer of 2015, Las Vegas was so flush with water that the city loaned some of its unneeded supply to help Los Angeles through a deep drought.[28]

Risks Remain

The great failure in Las Vegas water management is an odd one. Like many cities, it has repeatedly underestimated its customers' zeal for conservation, which results in overestimating how much water Las Vegas will need.

These failures are understandable. Water managers' incentives favor erring on the side of caution. The consequences of having too little water are far greater than the consequences of having too much. So Las Vegas has continued to pursue expensive and politically costly plans to import more water into the Valley from rural Nevada, water that the Valley's conservation success suggests may never be needed.

The economic downturn that began in 2007 and knocked the legs out from under the construction boom in the growing cities of the western United States has made Las Vegas water managers' job far easier. In 2009, the Southern Nevada Water Authority estimated that it would need enough water by 2014 to serve a population of 2.6 million people. But with growth stalled, the actual population in 2014 was just under 2.1 million.[29]

While the city's population grew more slowly than expected, Las Vegas residents' conservation performance also exceeded water managers' expectation. The agency's 2009 resource plan projected water use would drop to 240 gallons per person per day by 2015. But already by 2014, water use was down to 205 gallons per person per day.[30]

That is still a lot. Many other cities use less water per person. As we drove past strip malls and apartment complexes on the metro area's sprawling west side, Kurtis Hyde eagerly pointed out useless lawn borders and other overly wet landscaping that provided opportunities for more savings. The numbers bear this out. Water-use comparisons among cities are maddeningly difficult because of the differing assumptions made as each city calculates its own use. But by a reasonable apples-to-apples comparison, Las Vegas used 212 gallons per person per day in 2013, compared with Albuquerque, New Mexico's, 136.[31] Faced with extraordinary drought during the first decade of the twenty-first century, a quintet of major Australian cities demonstrated that it is possible to go far lower, averaging just 84 gallons per person per day.[32] Melbourne, the most extreme example, got all the way down to 65 gallons per person per day, one third as much as Las Vegas.[33]

City-to-city comparisons can be misleading. None of the Australian cities have as little natural rainfall as Las Vegas. Less rain means a need for more landscaping water if you want your city to be green. So we should not expect Las Vegas to cut its water use as far as Melbourne. Las Vegas officials also point to the fact that all of its sewage-treatment effluent is returned to Lake Mead for reuse, making its net water use far lower. But examples in places like Albuquerque, where sewage effluent also is reused, nevertheless suggest that, if Las Vegas chooses to continue to grow, it has room to significantly improve its water conservation and keep its water use within current levels. With some of the lowest-priced water in the western United States,[34] Las Vegas has not yet used what economists say is one of the most important water conservation tools of all—raising prices.

A return to the population growth rates of the 1990s and early 2000s could exhaust Las Vegas's current water-supply cushion. But the fundamental economic shift in the West's growth economy in the late years of the first decade of the 2000s suggests that this is unlikely. Economists expect a future with a much more modest growth rate for Las Vegas than the unsustainable boom years.[35]

Las Vegas leaders have been criticized for an unwillingness to constrain community growth in the face of possible water shortage.[36] But with savings from water conservation far exceeding demand from new growth, Las Vegas currently sits on a water supply cushion that keeps growing as the city stores unused water in aquifers for future use.[37]

Las Vegas is not alone in overestimating how much water its residents will need. In 2005, the Los Angeles Department of Water and Power projected it would need more than 700,000 acre-feet of water a year by 2015. The actual use in 2015 was less than 500,000.[38] In

1997, Albuquerque projected that it would need 125,000 acre-feet per year by 2015. By 2014, actual water usage was under 94,000 acre-feet and dropping faster than Albuquerque's population was growing.[39] This is a common pattern among municipal water systems in the western United States.

The Las Vegas region's political leadership has begun to recognize that residents are getting ahead of them in their attitudes toward water conservation. In September 2015, the Southern Nevada Water Authority revised its long-term water resources plan by using new assumptions that recognize increased conservation as well as the long-term slowing of the region's population growth. The new plan shows that, even in a worst-case water-supply scenario under which ongoing drought or climate change permanently depletes the Colorado River's flows, Las Vegas will have sufficient water supplies using its current portfolio until the 2040s.[40] But even as the new plan was being finalized, water users' conservation efforts continued to race ahead of the agency's projections, allowing Las Vegas to use less water and bank more for the future, or prepare for a time when climate change reduces the Colorado River's flow.

Sooner or later Las Vegas will again face difficult choices. Does it want to limit growth? Will it pursue new supplies, perhaps by building a controversial pipeline to pump water from groundwater basins in rural Nevada? Will it buy and transfer agricultural water rights from elsewhere in the Colorado River Basin, something not currently allowed but which might someday be possible as water management rules evolve? Or will Las Vegas residents choose what has thus far been the path of least resistance—simply conserving more? There is evidence that residents are already choosing the conservation path.

Whether that can continue is an open question. The important point is that Las Vegas has demonstrated as a community the institutional capacity to choose its path. The water-management structure built in response to the crisis of the late 1980s has proven robust enough to manage the water problems of Las Vegas to date. The decision of whether to continue growing is up to the community of Las Vegas, but if that is the path it wants to take, the success of the Southern Nevada Water Authority suggests that a shortage of water need not be a constraint.

The fountains at the Bellagio Hotel are unlikely to go dry.

CHAPTER 4

Negotiating the Rapids

Sɪᴅ Wɪʟsᴏɴ, ᴏɴᴇ ᴏғ Aʀɪᴢᴏɴᴀ's senior water managers, thought Jennifer Pitt was some sort of crazy tree hugger. Pitt, who worked on Colorado River issues for the Environmental Defense Fund, imagined Wilson as something akin to Genghis Khan. But until they set out on a boat trip together down the Colorado River through the Grand Canyon in the spring of 2004, the two had never actually met. The first night, Wilson, standing on a sandbar, kicked off the relationship by explaining why he didn't trust environmentalists.

It's remarkable what a river trip will do. There is something about nights in the depths of the Grand Canyon—the quiet water, the sliver of star-speckled sky between upraised cliffs, a beer shared on sandbars overlooking the river—that changes people. When Pitt and Wilson emerged from the canyon, the politics of Colorado River water management had changed with them. A fragile bond had been forged, one that would strengthen during the coming years into collaboration.[1]

"Maybe it was that we were wearing shorts and drinking beer, or maybe it was the magic of the river itself," Pitt later explained.[2]

The river trip, organized by one of the federal government's senior water managers, brought together federal officials, state water managers, and Pitt (the token environmentalist). There also was an air of theater about it. In the midst of growing drought, the feds invited five of the most prominent journalists covering water in the western United States.

The goal was to get the basin's key decision makers together in one place to talk about solutions to their shared problems. Bennett Raley, the Bush administration assistant interior secretary who organized the expedition, recognized that the issues were deep and that remedies handed down by the federal government were unlikely to work. Best to get the players onto the river, organize daily seminars, run some rapids, ply them with alcohol, and see what happened. "They will come up with a much more durable solution than we could by imposing one on them," Raley said.[3]

The trip on the Colorado River came at a pivotal moment. Lake Mead, full as recently as 1998, was dropping fast. In the spring of 2004, Mead was at its lowest levels since the early 1960s. Back then, the low water level was intentional: water managers had reduced flow into Mead in order to hold water back in order to fill Lake Powell, behind the newly built Glen Canyon Dam just upstream of the Grand Canyon. In 2004, on the other hand, the lake's levels were being pushed down by an increasingly alarming drought, combined with growing demands for water. Bad hydrology—low snowpacks and a warming climate—had struck at a time when growth in cities and farms left no slack. Water managers worried that in a world of increased demand and shrinking supply, they could not handle a shortage.

Nowhere was the situation more acute than at the bottom of the river, where the borders of Arizona, California, Sonora, and Baja meet. By the time the river's upstream users had taken their cut, the Colorado barely flowed past the last dam on the US-Mexico border. Environmentalists and the state of Arizona were reduced to fighting over trickles of salty water, flowing in a concrete channel into the mudflats of the Colorado River Delta. Relative to the larger uncertainties on the river, the fate of that salty supply didn't amount to much—perhaps 1 percent of the river's total flow. But as shortages loomed, the issue had become a flash point. If water managers couldn't solve this little problem, how could they tackle the big ones?

The disputed water, agricultural runoff too salt-laden to be much good for any human use, had flowed since the late 1970s across the US-Mexico border, past Yuma and San Luis, past US and Mexican farms and cities, and was dumped "unused" into the Santa Clara Slough, an old, dry Colorado River channel through the mud flats at the edge of the Sea of Cortez. But when it got there, something magical had happened. Add water and nature returned, and the water US managers intended to simply dump had become, by accident, the Cienega de Santa Clara, watering one of the most important wetland habitats in the region.

By accident, more than 40,000 acres of cattail marshes and open water had emerged from the mudflats. With much of the rest of the once-rich delta dried out by upstream water diversions, the accidental wetland had become the primary remaining habitat in the region for more than 200,000 migratory waterbirds, including 70 percent of the planet's entire population of the endangered Yuma clapper rail.[4]

On a river that typically carries 10–15 million or more acre-feet of water per year, the 114,000 acre-feet that had flowed down the Main Outlet Drain Extension to the Cienega de Santa Clara in 2003 is little more than a rounding error. But as Lake Mead dropped, Sid Wilson, general manager of the Central Arizona Project, worried about every drop. He was pushing the federal government to clean up the salty water and put it to human use.

This did not sit well with Pitt. Her employer, the Environmental Defense Fund, had earned a reputation as one of the environmental groups that the water managers could work with. It eschewed litigation, finding its greatest successes in developing collaborative relationships. When other people were suing to put water down the Colorado River's main channel through the delta in Mexico, Pitt was one of the leaders of an alternative approach.

But even in collaboration, Pitt and fellow environmentalists drew a line at the Cienega. Arizona's push to clean up that water and divert it for use upstream threatened the Cienega's precarious existence. Wilson and Pitt had taken to battling for the support of Arizona's newly elected governor, Janet Napolitano, and they were very publicly sparring in the press. In an *Arizona Republic* op-ed, Pitt and University of Arizona water law scholar Robert Glennon accused Wilson, by name, of an attitude "typical of Arizona's water past, with its single-minded dedication to increasing the supply of water."[5]

But until the two were invited on their trip down the Colorado through the Grand Canyon in the spring of 2004, they didn't know each other. By the time the trip was over, a change in attitude had taken place that rippled out through Colorado River Basin problem solving for years to come.

Wellton-Mohawk

The Cienega de Santa Clara is a beautiful outcome of one of the most dysfunctional episodes in Colorado River history, the effort to solve the dilemmas posed by farming the Wellton-Mohawk Valley along the Gila River in southwestern Arizona. The tale of the Wellton-Mohawk irrigation system and the Cienega de Santa Clara is byzantine, but understanding its details matters because it shows how important place-specific solutions become as we try to thread the needle of broader Colorado River Basin management.

Historian Evan Ward described the Wellton-Mohawk Valley as "an agricultural atoll in an arid sea of land."[6] Up the Gila River from the town of Yuma, Wellton-Mohawk is a narrow strip of green amid stark aridity, with brown hills rising sharply from either side of the irrigated valley floor. It is one of those places where you can almost feel the cocoon of safety created by irrigation. "The desert," a valley farmer once told me as we stood beside one of his water-rich vegetable fields, "is right over there."

Wellton-Mohawk farming began with one of the area's earliest attempts to use river water—the Mohawk Canal diverting water from the Gila River in the late 1800s. But the farmers were no match for the unruly Gila. When floods weren't washing out the irrigation headworks, drought left the farmers dry.

By the early 1900s, the farmers had turned to wells, forming the Antelope Irrigation District to run a steam-powered plant to pump water for the valley's growing farms. But by the 1920s, the area's agriculture had entered a period of decline because of a common problem—the buildup of salts in the groundwater.

All irrigation water has some level of salts and other minerals. When you spread it across a field, the plants transpire some of the water, concentrating the salts and minerals in the water that remains to soak down into the aquifer. Salinity is always a problem in irrigation agriculture. In Wellton, where the farmers kept pumping up water over and over again, the salt buildup eventually became intolerable.

To overcome the problem, the farmers turned to the federal government and the Colorado River. Work on the Wellton-Mohawk part of the water-delivery system did not begin until the 1940s, and by that time the salt problem had become worse. To solve it, the farmers and the federal government hatched a plan: they would bring in good, high-quality Colorado River water to replace the salty groundwater.

For a time, this worked. Farmed acreage, which had been declining because of the salinity problem, began expanding. But soon the Wellton-Mohawk farming district butted up against a familiar problem in Lower Colorado River agriculture. The bounty of imported river water, soaking down through the farm fields and into the aquifer, was raising the water table. To combat this, the farm community began installing a network of groundwater pumps, sucking up the salty groundwater and in effect replacing it with cleaner Colorado River water. But the salty water had to go somewhere. By 1961, that "somewhere" was a canal that carried it off and dumped it into the natural river channel draining the valley. That river channel drained to the Colorado, and the salty water ended up in Mexico, contaminating the water supply that the United States was delivering to Mexican farms and cities.[7]

The United States government, straight-faced, advanced a boldly cynical argument. The 1944 treaty between the two nations required that the United States deliver 1.5 million acre-feet of Colorado River water per year to Mexico. The treaty said nothing about the quality of that water. Usable or not, the salty Wellton-Mohawk water met our water-delivery obligation.

This did not sit well with farmers and cities in Mexico, as the Wellton-Mohawk drainage water was strangling Mexican crops and leaving municipal water taken from the Colorado River south of the border undrinkable. The Mexicans complained, and by the early 1960s the tension over the issue had grown from a regional water management problem into an international diplomatic conflict.

Negotiations dragged on until 1973, when the two countries finally came to terms on an agreement grandly titled the "Permanent and Definitive Solution to the International Problem of the Salinity of the Colorado River." The national governments agreed to enforceable salinity standards for the water delivered at the Mexican border, and the United States agreed to build a concrete channel to bypass the salty Wellton-Mohawk water all the way to the Santa Clara slough. The water delivered through that channel, because it was unusable, would not be counted against the Mexicans' share of the river. That is, the United States still had to deliver 1.5 million acre-feet per year, but the bypassed salty water would not be counted as part of that obligation. The US government would have to find another source.

Critically, the Salinity Control Act, the federal legislation that implemented the US terms of the deal, specified that the "lost" water being sent to Santa Clara would not come out of Arizona's share of the Colorado River distribution. It would be "a national obligation." The notion that Arizona and the Wellton-Mohawk farm community should account for their own water use did not seem to register with the planners. Rather, the decisions made in the 1970s assumed that US taxpayers, as well as water users across the Colorado River Basin, would bear the burden of Wellton-Mohawk's lousy hydrology.

A "National Obligation"

To fulfill this "national obligation," the Salinity Control Act prescribed old-school engineering—US taxpayers would fund a $256 million desalination plant on the US-Mexico border to treat the saline water coming out of Wellton-Mohawk and return the cleaned-up water to the Colorado River for use in Mexico. This followed the longstanding US pattern of spending a lot of money on infrastructure for water for human use.

In the interim, while the Yuma Desalting Plant was being built, the federal government extended the Wellton-Mohawk drain into Mexico, where it began the accidental creation of the Cienega de Santa Clara. To make up the difference until the desalting plant was completed—a hundred thousand acre-feet or more of clean Colorado River water had to be delivered annually at the US-Mexico border to make up for the salty water being bypassed to the Cienega—the US government simply released more water from Lake Mead.

In 1992, the Yuma Desalting Plant was completed and put into operation, but it was quickly shut down when a flood from the Gila River tore out its intakes. At that point, bureaucratic inertia set in. Lake Mead had plenty of water, no one was worried about shortages, and it was cheaper and easier to just cover the shortfall with extra releases from Lake Mead rather than spend the money and energy to get the Yuma Desalting Plant running again. Surplus water from Lake Mead would be released for Mexico, and the salty water could be left to the Cienega.

With plenty of water in Lake Mead, taking a bit out of the giant reservoir each year to avoid the financial cost of running the desalting plant as well as the environmental and cultural cost of drying up the Cienega made sense. Thus, the happy accident of the Cienega de Santa Clara, one of the only remaining large tracts of nature in the otherwise dewatered Colorado River Delta, endured. This was one more luxury afforded by the surpluses flowing down the river in the 1990s, but also one more problem deferred. As we see time and again in the Colorado Basin, when there is plenty of water, it is far easier to just put things off.

But by the early 2000s, as drought set in, that luxury was disappearing. Political deals made in the 1960s left Arizona last in line for water during times of shortage. Feeling vulnerable because its water would be the first to be cut as the reservoirs upstream dropped, Arizona wanted the federal government to step in, halt the flow of water to the Cienega and run the water through the Yuma plant to clean it up and deliver it to Mexico, thus helping slow the decline of Lake Mead.

The Yuma Desalter Working Group

That was the state of things when Bennett Raley invited environmentalist Jennifer Pitt and Arizona water manager Sid Wilson to raft the Colorado. By the time the trip was done the pair had laid the foundations for an informal working group to try to bridge the gap between the two sides. In the year that followed, the group's members came up with a report that stands as an odd but remarkably important document in the efforts to sort out the Colorado River Basin's complex problems.

Never before had such a diverse group of Colorado River stakehold-

ers—representatives of federal and state governments as well as environmentalists—come together. The group represented an impressive body
of expertise on the river's water-management issues, and the individual
members were carefully chosen. In order to work, this had to be com-
posed of people who got along well.[8]

The "fundamental objectives" laid out by the working group sound
crazily optimistic: reduce the risk of shortage to Lower Colorado River
Basin water users while at the same time maintaining the wildlife and
habitat at the Cienega de Santa Clara. And also, "to the extent possible,"
improve water quality for cities in the border area. As if that weren't
ambitious enough, the solution shouldn't cost very much money. Plus,
it should be something that could be done quickly. The whole thing
amounted to a Christmas list.

In bringing the people together, Wilson and Pitt laid out an important rule. Members of the working group "were asked to participate in
the process as individuals rather than as stakeholders. In other words,
members did not have to represent the position of their employers nor
in any way was it assumed that the groups or agencies they normally
represent would even agree with or endorse the Workgroup's recommendations."[9] The caveats in the report's footnotes are almost comical
in the pains they take to point out that the results "do not represent the
official position of, or endorsement of" the various agencies represented.

The process has been criticized for lack of official participation, and
therefore the lack of binding commitments, but in fact that was one
of its strengths. As we will see later, scholars who study collaborative
efforts to solve common-pool resource problems have found that such
informal conversations can set the stage for later deals. Precisely because
there are fewer constraints and the stakes are lower, participants get a
chance to really understand one another's perspectives. Elinor Ostrom,
the political scientist and Nobel laureate who pioneered research into
this approach, called it "cheap talk."[10]

In the short run, the relative informality meant that there was no concrete, institutionalized way to carry out the working group's recommendations. But in the long run, the forum proved profoundly important. It is not that the ideas behind the recommendations were new. But the workgroup created space for a conversation that had never before been possible—about how those ideas could fit into broad solutions. For the first time, Arizona water managers acknowledged the importance of the Cienega, and they conceded that any solutions to Arizona's water shortfalls couldn't simply brush the accidental wetland away. Environmentalists, in turn, acknowledged that Arizona's vulnerability to shortage was real, and that any solutions to the river's environmental problems needed to recognize that reality. In the decade that followed, many of the ideas worked out in the Yuma-Cienega meetings shifted from "cheap talk" to official dialogue.

Among these ideas was the construction of a new storage reservoir to capture excess flows on the lower river that frequently happened during rainstorms, when farmers couldn't use water they had ordered from Lake Mead. The workgroup also suggested water-conservation steps on the Mexican side of the border, to which the wealthier United States would contribute. The group recommended finding a way to make use of excess groundwater in the Yuma area, a suggestion that remains a central option among decision makers today. And the group recommended managing flows to the Cienega as an environmentally valuable phenomenon, rather than simply an accident of history. Once heresy for water managers, this concept is now a given in Colorado River Basin water policy discussions.

Even more significantly, the workgroup recommended a "pilot, Basin-wide, consumptive-use reduction and forbearance program." This was revolutionary. Up to that point, water conservation programs had narrow, specific purposes: save water over here so that we can use it over there. But the Yuma group was recommending something different—conserving water in order to simply leave it in the river or the basin's reservoirs, in trust for all. The group's report led to a pilot test of the desalting plant to gain information about its technical operation and the costs of its future use, something that before the detente would almost have certainly led to conflict between environmentalists and the water-management community.

Most importantly, the working group's recommendations provided a framework for formal, multiparty negotiations that eventually expanded to include the Mexican government and that nation's water agencies and environmental groups. These talks led to the historic Minute 319 US-Mexico agreement that for the first time intentionally released environmental river flows back into the Colorado River Delta.[11] It was Minute 319's environmental flow that drew me to the San Luis Bridge in March 2014. But the historic agreement, rooted in large part in the Yuma-Cienega discussions, had a far broader reach in that it began to deal with important cross-border environmental and water-management issues.

All of the successes that grew out of the Yuma working group are place-specific. They depended on active participation by experts in the unique geographies where the water had to be managed. So the specific solutions do not generalize to other parts of the Colorado River Basin, or to other challenges in water management. What does generalize is the method, the ability to break down barriers, starting with "cheap talk." What began on a Grand Canyon river trip laid the groundwork for real cooperation between environmentalists and water managers, poking holes in the myth of inevitable conflict and creating a model that can be replicated across the West.

CHAPTER 5

Arizona's Worst Enemy

IN THE STRUGGLE TO SHARE THE Colorado River, Arizona has always been its own worst enemy. From the time statehood was established early in the twentieth century, Arizona politicians viewed the Colorado River flowing down their western border as the key to the desert state's economic future. But their approach was always pugnacious rather than collaborative. The state refused to sign the Colorado River Compact, the basin's first grand water-sharing deal; it took its neighbors to court again and again; and Arizona once went so far as to dispatch its National Guard in a "war" against California and the federal government over water.

Arizona's behavior is driven by psychology as much as geography. Life in the desert can be harsh, stifling. In the aptly nicknamed "Valley of the Sun," the Phoenix metro area where nearly two of every three Arizonans live, you can feel the summer heat press in around you. Water, especially imported water pumped up from the Colorado River to the west, has always offered a cocoon of cool comfort. Those palm trees, pools, and lawns in the cities of central Arizona provide more emotional than economic benefit, but they come at the expense of significant water use.[1]

Fountain Hills, outside Phoenix, Arizona (© John Fleck).

Along with an unforgiving landscape, Arizona has long had an inferiority complex that dominates its relationship with California, its bigger, richer, more-populous neighbor. When it comes to water, the complex has played out in a century-long political psychodrama of fear that Californians were out to steal all the Colorado River's water. Arizona's unwillingness to collaborate meant it was largely left out as the Colorado River's big water-distribution projects were being built during the first half of the twentieth century. When Arizona finally got its big canal, its decades of intransigence left it with little bargaining power, forcing it to accept a deal on unfavorable terms that have left the state's water supply vulnerable as the river runs short.

The irony is that within the state's boundaries, Arizona has done a far better water-management job than it is frequently given credit for, gracefully handling the reduction in desert agriculture that so many argue is essential for sustainable water use in the Colorado River Basin.

That and important groundwater regulations have made the populous southern two-thirds of the state the largest major groundwater-using area in the West where scientists have actually documented aquifers on the rise after decades of over-pumping. Imported water from the Colorado River helped, but reductions in groundwater pumping played an even bigger role, such that the state's total water use peaked around 1980 and has been declining ever since.[2]

But Arizona's pugnacious style keeps surfacing, jeopardizing the cooperation needed to reduce water shortages across the basin. Arizona still seems to believe the old canard that "water's for fightin' over."

The Valley of the Sun

Standing on the roof of Phoenix's old Verde River water treatment plant, you can readily see how an uneasy relationship with water has defined life in this desert community. A ribbon of green, the aptly named Verde River, snakes through tan hills staked with saguaro cactus. Just as the Verde pinches between hills one last time before flowing out into the valley, the treatment plant scoops up the river's precious flows, treating the water and putting it to what in the lingo of the West has always been called "beneficial use."

Kathryn Sorensen, head of the city's water department, took me to the roof to explain just how water moves through the Phoenix valley. An Arizona native and economist by training, Sorensen is responsible for making sure 1.5 million people in the desert get water when they turn on the tap. The greater Phoenix area grabbed its native rivers, the Salt and the Verde, early on. Look at the old pictures and you will see an incongruous sight—vast acres of cotton, an empire of irrigated agriculture on the desert floor. But from the beginning Phoenix needed groundwater to supplement the supplies it got from the Salt and the Verde Rivers, and everyone knew that in the long run that was unsustainable. Phoenix always yearned for more, looking with longing and anxiety off to the west at the Colorado River. Throughout the twentieth century, ground-water depletion accelerated as the region's farms and cities grew.[3] They were mining "fossil water" left in the region's valleys of sands and gravels over eons, and they were taking it out far faster than nature could replace it. "It doesn't recharge at a very fast rate," Sorensen explained.[4]

Arizonans watching their population grow have always yearned to wean themselves from groundwater, and a fresh supply of water imported from the Colorado River long seemed the only answer; the alternative, they feared, was doom. A cartoon from the 1940s shows the grim reaper scratching "NO WATER" with his bony fingers across a map of Arizona, with the fat, wide Colorado River flowing untouched down the state's western edge. "Without water," the cartoon's caption read, "Arizona's economy will perish."[5]

The native landscape here has always felt like destiny. Back in 1970, a pair of young economists, William Martin and Bob Young, wrote that Arizonans believed that the "stark vistas of our desert surround-ings" offered "self-evident proof" that water was the region's limiting resource. Martin and Young disagreed. Arguing, as economists do, from data, they showed that Arizona's economy was doing better even as water grew scarcer.[6] But whether or not water controlled the state's economy was beside the point. Arizonans have always believed it to be true. Arizona's first grand dam was not even complete when delegates to the state's 1910 constitutional convention enshrined their vision of the future in the state seal: ". . . at the right side of the range of mountains there shall be a storage reservoir and a dam, below which in the middle distance are irrigated fields and orchards reaching into the foreground, at the right of which are cattle grazing. . . ."[7] The seal embodied Arizona's ambition for water and Arizonans have feared its lack ever since.

Throughout the twentieth century, it was an unquestioned article of faith that central Arizona needed more water, and that the Colorado River was the place to get it.[8] Arizona's greatest politician, the swaggering county sheriff turned congressman turned senator Carl Hayden, built a career on the drive for Colorado River water, and the battle over how best to get it defined the state's politics for half a century. Water was "fuel for growth."[9]

The Skinny Kid

In Colorado River water politics, Arizona has always played the skinny kid, bullied by a burly California neighbor bent on kicking sand in its face and stealing its water. The "victim" narrative was often false, but what mattered is that Arizona believed it. The state's inferiority complex has played a central role in the Colorado River Basin's water politics for a century, creating a pattern of obstruction that delayed the very water projects Arizona said it needed and a history of conflict that continues to this day.

Arizona's pugilistic stance was summed up in the 1920s in a comment by Phoenix resident Ralph Murphy, as Congress considered legislation to enable the great concrete plumbing needed to move the river's water across the West. Murphy described the bill as many Arizonans saw it—as a scheme to deprive his state of water that could otherwise be used to irrigate Arizona farms, sending it to California instead: "I believe Arizona will fight this scheme to the last ditch."[10] Never mind that the system of federally subsidized dams that Arizona was fighting to stop were the same dams it would ultimately need to corral its share of the river. For decades, Arizona lived up to Murphy's pledge.

From the beginning, Arizona felt outmatched. In 1912, at statehood, it counted a population of just 217,000 to California's 2.7 million.[11] Residents saw a raw desert with little going for it other than the waters of the Colorado River flowing down its western border.

That water should rightfully be ours, Arizona state senator Fred T. Colter argued. Arizona had "one-half of the entire drainage area of the Colorado River Basin" and was, Colter contended, entitled to half of the water. If Colter's geography was a little bit wrong (Arizona did not make up half the basin, though it was close), his legal analysis was in left field (no law allocated a river's water by basin acreage). But the idea that Arizona was being bullied into giving up its hydrologic birthright resonated. Arizona, sparsely populated, growing slowly, newly admitted to the union, was "a baby state" set upon by powerful neighbors, Colter charged.[12]

When Congress appeared ready to push its way forward without Arizona's consent, Hayden thundered on the floor of the Senate that the waters of the Colorado were being unjustly divided "primarily for the benefit of California." Hayden and Arizona senator Henry Ashurst launched a filibuster that delayed action for a time, while nearly bringing senators to blows, but in the end the Arizonans lost.[13]

Arizona was never entirely clear about what it needed the water for beyond fighting off the stark vistas of its desert surroundings. In the early years, it argued that the water was needed to prevent the collapse of its agricultural economy, but this gradually shifted until, by the 1960s, the water was needed for the great cities growing in the state's central valleys. Whatever the reasons, the perpetual need for more water became Arizona political dogma, and Arizonans followed Colter's charge.[14] Their effort not only failed but, for most of the twentieth century, had the opposite effect. Arizona's neighbors grew, as did their use of the Colorado River. Arizona meanwhile painted itself into a political corner that prevented it from getting the water it so desperately wanted.

Fighting wasn't working, but Arizona seemed to know no other way.

The Arizona Navy

When the seven Colorado River Basin states negotiated the 1922 Colorado River Compact, an agreement to divide the river's water, Arizona was at the table and agreed to the deal. It was the first uneasy attempt at sharing the Colorado River's water rather than fighting over it. But when Arizona's negotiator, Winfield S. Norviel, took the deal home for ratification, the state's political establishment balked. Arizona's political leadership changed in the November 1922 election, and newly elected Governor George Hunt feared that the agreement locked in California's advantage, leaving his own state uncertain about future access to the water. At every turn, Arizona opposed the ensuing legal and legislative steps needed to develop the Colorado River's water.

In 1928, Hayden and other Arizona politicians fought a losing battle in Congress against the Boulder Canyon Project Act, the federal legislation that authorized construction of what would become Hoover Dam, along with the infrastructure needed to move the Colorado River's water to California. The legislation was the next step needed to move the agreement embodied in the 1922 Compact forward. But Arizona's refusal to ratify the Compact had left the state out of the ensuing distribution of the political pork as the necessary implementing legislation was being drawn up. The other states went ahead without Arizona and crafted the 1928 legislation needed to build Hoover Dam. Despite Arizona's intransigence, the bill included provisions to ensure that a share of the water was allocated to Arizona, as the 1922 Compact had decreed. But the 1928 legislation included no federal money to build the infrastructure needed to use it.

Robbed of the chance to develop a share of the river's water, Arizona resorted instead to fighting with its California neighbors over theirs. In the fall of 1934, Arizona's struggle with neighboring California came the closest the West has ever come to a literal war over water.

On November 12, Arizona National Guard Major Franklin Pomeroy sent an urgent telegram to Joe Bush of Parker, Arizona: "WILL REQUIRE USE OF YOUR SMALL BOAT COMMENCING WEDNESDAY."[15] Bush and his wife, Nellie, ran a ferry service near the Colorado River's junction with the Bill Williams River, close to the future site of the Parker Dam. By the mid-1930s, they were busy toting survey crews from the Metropolitan Water District of Southern California back and forth across the river.

Parker Dam was to be built on the Colorado River some 150 river miles downstream from Hoover Dam, then under construction on the Arizona-Nevada border. Hoover would store huge volumes of the river's big flows, then pass them downstream as needed to Parker Dam, creating a reservoir that would be the diversion point for Southern California's Colorado River Aqueduct water. Parker Dam's west-bank abutment would be in California, its east-bank abutment in Arizona. On their side of the river, Californians were already at work on the 242-mile Colorado River aqueduct, which would carry Colorado River water to the rapidly growing Los Angeles Basin.

While the state had played a losing game in Congress and the courts, the physical reality of Parker Dam's eastern anchor on Arizona soil gave the state some leverage. Arizona governor Benjamin Moeur ("a showman politician in the grand carnival style," in journalist Marc Reisner's words) dispatched 100 National Guardsmen to block construction work on the dam's eastern footings.[16]

Nellie Bush flew the Arizona flag on her ferry boat, the *Julia B*—the sole vessel in what a Los Angeles newspaper laughingly called "the Arizona Navy." The *Los Angeles Herald-Express* documented the affair with a half-page cartoon in the manner of a military campaign map, with cowboy-hatted politicians storming out of the state capital in Phoenix declaring, "I tell you we've been invaded! Violated th' sovereign rights of our state, it's war!"[17]

Arizona deserved to be mocked. No one was going to shoot anyone over the question of whether a dam could be built at Parker. Yet while Major Pomeroy's feeble armada was more theater than a genuine call to arms, it established a precedent. Conflict had become the norm. More often, the battles were waged in the halls of Congress and in the courts, but the results were the same. After a promising early twentieth-century effort to share water, Arizona decided that it didn't like the results of the process and turned away from collaboration. The ongoing clashes damaged efforts to manage the Colorado River, with consequences that linger today. Yet the biggest harm came to Arizona itself, which learned far too late that when it went to the battlements, it lost more often than it won.

Congress resolved the Parker Dam standoff the following August with the Rivers and Harbors Act of 1935, which swept away Arizona's objections and cleared the way for the dam's construction. Within a month, President Franklin Delano Roosevelt was standing 150 miles upstream atop the nearly completed Hoover Dam to, in his words, "celebrate the completion of the greatest dam in the world." Despite the dam's eastern footings being anchored in his state's bedrock, Arizona governor Moeur did not attend the ceremony.[18]

Preparing for Litigation

In the summer of 1944, Arizona's water managers knew they were living on unsustainable groundwater and borrowed time. Spread along the Gila River southwest of Phoenix, the Roosevelt Irrigation District began drilling wells in 1928. Just fifteen years later, its pumping had more than doubled, and its aquifer was dropping fast.

"This water table has dropped from sixteen to forty feet," district superintendent John P. Van Denbergh told members of the US Senate's Committee on Irrigation and Reclamation, who had gathered in Phoenix to hear about Arizona's water needs. Ten of the agency's ninety wells had already been abandoned. Van Denbergh's drainage engineers figured that the district needed to cut its pumping nearly in half to bring demand into balance with supply.[19]

All across Arizona, water development was following the same path. The solution, the state's leaders thought, lay in finding a way to move some of Arizona's share of Colorado River water into central Arizona.

Arizona senator Carl Hayden listened quietly as, first, representatives of the US Bureau of Reclamation and then a parade of witnesses including Van Denbergh made the case for a major federal effort to bring Colorado River water to arid central Arizona.

The federal government seemed willing to oblige. "Since the time of Alexander the Great," Bureau of Reclamation commissioner Harry Bashore told the committee in some of the most grandiose rhetoric to come out of a grandiose agency, "men have complained that they could find no more lands to conquer. We of the Reclamation Service [by then officially renamed the Bureau of Reclamation] know that there are still lands of the great Southwest where our two weapons—water and power—can conquer drought and despair. And as these enemies of mankind are routed, we can build a greater Southwest with the help of those who will seek employment here in our public works."[20]

But by now Southern California's urban-suburban empire was booming, built with imported Colorado River water as its lifeblood. Its voracious appetite for the river's water had exposed ambiguities in the language of the Colorado River Compact and the ensuing federal legislation authorizing construction of Hoover Dam.

Whether the situation would have been clarified from the start if Arizona had ratified the compact and joined its neighbors in development of the Boulder Canyon Project Act we will never know. But by 1944, the lack of clarity seemed an insurmountable obstacle to Arizona's water dreams. The Bureau of Reclamation and its partners in Arizona could plan all they wanted, and the bureau was happy to do Hayden's bidding on this score. But Arizona's refusal to ratify the Colorado River Compact, and its subsequent decision to "fight to the last ditch," had left a cloud of uncertainty over the question of how to divide the Lower Basin's share of water allocated under the Colorado River Compact. And without an answer to that question, it was difficult to proceed with any project. The only answer was litigation, it seemed to E. B. Debler, the bureau's director of project planning. "A reliable division of waters may require a decision by the Supreme Court of the United States," Debler told Hayden's committee.[21]

Climbing the Courthouse Steps

The legal drama when one US state sues another is unique. States are sovereign entities, and the only venue with the jurisdiction to sort out their complaints is the United States Supreme Court. Whereas, in most legal disputes, the Supreme Court is the court of last resort, in interstate litigation, it is the first and only. Lawyers call it "the court of original jurisdiction."

Thus it was that J. H. "Hub" Moeur, chief counsel for the Arizona Interstate Stream Commission, walked out of Carl Hayden's Washington, DC, office in the summer of 1952, armed with a legal brief. As the nephew of Governor B. B. Moeur, who two decades earlier had used the National Guard to declare a war of sorts on California, Hub Moeur had a family history of fighting over water. With Hayden at his side, he climbed the steps to the US Supreme Court building and delivered Arizona's petition in what would become one of the court's most consuming cases.[22]

Moeur's argument was simple. The Law of the River limited California to 4.4 million acre-feet per year of Colorado River water. That constraint notwithstanding, California and its water agencies had signed contracts with the Bureau of Reclamation for nearly 5.4 million acre-feet of water. More importantly, Moeur alleged, California and its water agencies "have caused the construction of works of a capacity to divert more than 8 million acre-feet annually." Arizona feared that future. His state, Moeur wrote, needed 3.8 million acre-feet per year of water from the river "in order to sustain its existing economy."[23] Indeed, Arizona had already taken the steps to begin using that water, laying out the legal and physical route for the Central Arizona Project, with land staked out for what at the time was called the Granite Reef Aqueduct, a system of pumping plants and a canal to carry Colorado River water to the Phoenix valley. This was the latest version of a plan that had existed in various forms since the 1920s. But California, which had begun using the water Arizona could not, stood in the way.

The brief by Moeur and his colleagues oozed with the sense of hurt that Arizona had felt over the previous three decades, cataloguing the injustices heaped upon it by the way its California neighbors and the federal government were developing the Colorado River's water. To read Moeur's words today is to understand Arizona's anger in looking across the river and seeing those two big aqueducts carrying water off to California.

As recently as 1946, California had only taken 3.4 million acre-feet, but in the years since then, California's take had risen steadily. By 1951, the state had taken 4.5 million acre-feet, finally passing the magic "4.4" that Arizona thought should be California's limit. As Moeur and Hayden filed their case in the summer of 1952, the number continued to rise. "At the rate diversions have been made during the calendar year to date," Arizona's lawyers wrote, California "will divert at least 5.43 million acre-feet."

This was Colorado River brinksmanship in full display. Knowing that Arizona was preparing to take the case to the Supreme Court, California began cranking up its diversions, especially to the vast desert farming area of the Imperial Valley in the state's southeast corner, hoping to lock in a higher number.[24]

For Arizonans, Moeur and his colleagues argued, much was at stake. They implicitly acknowledged the state's central problem, as so clearly outlined in the testimony before the Senate Committee on Irrigation and Reclamation at that Phoenix hearing eight years earlier: Arizona was living beyond its water means.

"Arizona," they wrote, "is an arid state. Irrigation is essential to its successful agriculture, and much water is needed for domestic, municipal, and industrial purposes. Precipitation is insufficient to satisfy the need for water."

The Decision

The headline on the front page of the June 4, 1963, *Arizona Republic* crowed: "Arizona Wins Water Suit; $1 Billion Project Next." The sub-head read like a victory dance: "California Loses by 5–3." After a decade before the Supreme Court, Arizona appeared to have emerged victorious.

The newspaper framed the Supreme Court's ruling in the case of *Arizona v. California* as "a personal triumph for Sen. Carl Hayden." Indulging in some decidedly partisan boosterism, the *Republic* wrote, "The court adopted the apportionment laid down by Hayden with the passage of the 1928 Boulder Canyon Project Act."

In fact, Hayden had tried and failed to curtail California's allotment under the 1928 law, filibustered the bill, and then voted against it. But while wrong on the specifics, Arizona's public narrative was right in spirit. The Supreme Court's decision that day looked like an enormous victory for both the state and Hayden.

On the case's central points, the court had given Hayden and Arizona what they had been asking for the whole time. Yes, Arizona was entitled to the 2.8 million acre-feet of water per year of water from the main stem of the Colorado River that it wanted, plus the right to use tributary water from the Gila River without having that count against Arizona's share. California would be held to 4.4 million acre-feet. California could use surplus if it was available, but not if Arizona needed the water.

"Arizona gave a sigh of relief and joy yesterday that has been pent up for forty years in the battle with California over the Colorado River," the *Republic* said. Tucked in the bottom-right corner of the *Republic*'s front page was the only other mention of Arizona's archrival: "California's Opposition Will Move to Congress." This fight was not over yet.

But as a matter of water arithmetic, the ruling compounded the mistakes of 1922, amplifying the problem that the Law of the River had allocated more water on paper than the real river could provide.

The CAP

Carl Hayden wasted little time, but it still took years for Arizona to get its Central Arizona Project. The day after the ruling was handed down, he and the other four members of the state's congressional delegation introduced legislation in both the House and the Senate authorizing construction of the Central Arizona Project "for the purposes of furnishing supplemental irrigation water and municipal water supplies to the water-deficient areas of Arizona and western New Mexico, through the direct diversion or exchange of water."[25]

Hayden had been in Congress for more than half a century by that time, with water projects for his rapidly growing desert state always at the center of his legislative agenda. He had "always figured" that he'd be around to see the Central Arizona Project built; he just had not expected it to take so long. "That's the main thing I have left to do," Hayden said at the time.[26]

As had been the case in 1952 when Arizona filed suit against California, the notion that "water-deficient" was the product of *both* supply and demand seems still not to have occurred to the Arizonans, or at least if it did they kept their mouths shut about it. In the years since the Bureau of Reclamation had begun looking at the feasibility of a Central Arizona Project in the late 1940s, the state's population had doubled, to 1.4 million people. The land under irrigation had risen to more than a million acres, and the bureau found that groundwater levels in the Phoenix and Tucson areas were dropping at ten to twenty feet per year in some places.[27] Already, according to the Bureau, Arizonans were over-pumping their aquifers by 2.2 million acre-feet per year, far more than even a completed Central Arizona Project could deliver. This was no secret at the time. "Arizona cannot make up this deficit," explained a

Congressional Quarterly summary of the situation, published in Arizona newspapers in June 1963.[28] The notion of using less seems never to have crossed Arizona's mind.

The general plans for the Central Arizona Project (CAP) had been in hand for decades. Even after the Supreme Court's ruling clarified water entitlements, it nevertheless took five years and a huge concession by Arizona to finally win congressional approval for the Central Arizona Project.

California used its congressional muscle to claw back a piece of what it had lost in the courts, achieving a political deal with ramifications that still echo today. To get the CAP agreement through Congress, Arizona agreed that its Central Arizona Project supplies would be behind California in the priority line if there was ever a shortfall on the Lower Colorado. In essence, Arizona was once again flummoxed by California's rapid development and the old doctrine of prior appropriation. California would be limited to 4.4 million acre-feet, but would stay first in line for the Colorado River's water if the river got so low that there was not enough for that plus Arizona's newly allocated share.[29]

Given the reality of the math, it seems odd that Arizona would have accepted such a deal. During the hard-fought debates before the US Supreme Court, it had become clear that the only way the three Lower Basin states would get their full 7.5 million acre-feet per year was if there was surplus in the system—unused Upper Basin water flowing south. Wouldn't that mean that, once the Upper Basin finally developed its own water use, Arizona would then see its Central Arizona Project canal run dry? The answer to that question lies in what Arizonans believe was a promise that accompanied the compromise: that the shortage would be solved by "augmentation" of the Colorado River. It was a time of dreamy engineering.

In a December 1967 speech in California, a few months before the final legislation passed, Arizona representative Morris Udall laid out what he saw as the terms of the deal: an aqueduct to Phoenix and Tucson, an agreement that California would get priority for its 4.4 million acre-feet, and a commitment to augmentation: "through desalting, weather modification, and Lower Basin salvage and conservation, especially through prevention of waste by California irrigation districts."[30]

With that deal, the Central Arizona Project was legislatively launched in 1968. It would take another twenty-five years for water to actually reach its southeastern terminus near Tucson, and when it did, the project's completion fundamentally changed the arithmetic of Colorado River water.

Cotton Dethroned

When we think today about Arizona's water problems, we imagine large lawns in sprawling suburbs in and around Phoenix, golf courses, and "misters"—those devices that fritter away water into the hot desert air to cool the customers eating at outdoor restaurants in the Valley of the Sun.

But to really understand Arizona's water problems, and what their solutions can look like, you have to start with cotton. In the 1950s, as Arizona was battling in the courts over who was entitled to how much Colorado River water, it was agriculture, not growing cities, that drove Arizona's persistent desire for water. And it was cotton, a thirsty crop that thrived in the state's central valleys *if* farmers could find enough water, that drove Arizona agriculture.

Arizona's cotton explosion in the 1950s was propelled by two things. The first was cheap groundwater that allowed farmers to expand their acreage. Farmers could essentially pump as much as they wanted or could afford. The second was price. From 1940 to 1950, the price that Arizona farmers received for a pound of cotton more than tripled, and the amount of acreage planted in cotton and the amount of water used to irrigate it exploded. In 1940, there were 221,000 acres of cotton in Arizona; by 1953, that had jumped to 695,000 acres. Farmers were turning desert land and cheap pumped groundwater into cash, and quickly.

As with much of Colorado River Basin agriculture, alfalfa had always been a big part of Arizona's agricultural portfolio. But over the course of the 1950s, cotton came to dominate the state's agricultural economy. In 1953, the peak year for cotton acreage in Arizona, three times as much land was devoted to cotton as was planted in alfalfa. But the revenue gap was even bigger than the acreage gap—$178 million in cotton sales that year, more than ten times the revenue from Arizona alfalfa sales.[31]

In 1959, when county extension agent George W. Campbell Jr. surveyed farming in Pinal County, the stretch of desert valley between Phoenix and Tucson that was one of the most important agricultural regions in the state, alfalfa was barely an afterthought. Cotton was king. "During the peak of the cotton harvest," Campbell wrote, "all who wish to be are employed."[32]

But Arizona's groundwater regulation was a tangled mess, which placed few restrictions on pumping, and, as a result, aquifers were rapidly dropping across the state. To the federal government, this was a recipe for disaster, because imported Central Arizona Project water was likely to be more expensive than groundwater. If anything, Colorado River water would be added to central Arizona's water portfolio, while excessive groundwater pumping to feed cotton and other agriculture would continue. Colorado River water would not untangle the mess.

The 1968 law authorizing the Central Arizona Project included language intended to prevent the use of the imported water to simply expand agricultural acreage while the state's farmers continued to over-pump. The law mandated that, for every contract for delivery of CAP water, "there be in effect measures, adequate in the judgment of the Secretary, to control expansion of irrigation from aquifers affected by irrigation in the contract service area."[33]

On a visit to Phoenix in 1979, Interior secretary Cecil Andrus made clear to Arizona's political leadership that if they didn't come up with a binding, enforceable groundwater-management regime, there would be no CAP. Andrus's comments were a big deal. The federal government had a long history in the West of deferring to state preferences when it came to water-rights management. The Bureau of Reclamation, for example, had long either formally or informally looked the other way when it came to enforcing the early legal limitation restricting federal irrigation water to farms 160 acres in size or smaller. So while the Colorado River Basin Project Act of 1968 had included a technical requirement for groundwater management, Andrus's sharp and public comments reflected a shift on the part of the federal government. Arizona was going to have to get its water-management house in order if it was going to get the help of US taxpayers to build the Colorado River supply system it so desperately wanted.

Within Arizona, the political dynamics were complex. In much of the development of the West's water, agricultural interests had dominated, invoking their prior-appropriation water rights and economic power to lock up water supplies. But Arizona was coming late to the scene. Never before had a water project this large been built at a moment in history when municipal interests in a state were this powerful. Agriculture was no longer the economic powerhouse in Arizona that it had been in other regions when they were developing their water projects. By the late 1970s, the cluster of cities that made up the greater Phoenix metro area, not Arizona agriculture, had come to a dominant position in Arizona politics. And the mining industry, important to Arizona's economy, also played a critical role. Mining needed water, and miners and agriculture had clashed in the past over groundwater pumping rules, which had already created an alliance between mining and municipal water agencies. In 1979 and '80, as Arizona's political establishment wrestled with how to respond to Andrus's pressure, the mining-municipal alliance proved critical.

It was clear to everyone that a reduction in agriculture was the only way to accomplish the groundwater regulation the state needed. Cotton farming had ebbed in the years since its 1950s boom, but by the late 1970s it was back in a big way, leading to huge groundwater overdrafts. In the negotiations over how to respond to the new federal requirement, farmers pushed hard for a buyout: pay farmers to reduce their irrigated acreage. Mining and municipal interests pushed back.

For six months, a negotiating team representing state leadership along with the core constituencies—mining, agriculture, and municipalities—met "informally" in an effort to sidestep the glare of Arizona's open-meetings law, and tried to work out a deal.[34] The final agreement placed groundwater-management restrictions on five "Active Management Areas" in the state, most importantly including the Phoenix and Tucson metro areas. There, existing pumping was grandfathered in, with provisions added to restrict future expansion of agriculture and to require municipal growth to demonstrate an "adequate supply" for the next 100 years. Over time, pumping would have to be decreased. Tweaks in ensuing years weakened the law's provisions, allowing growth with a promise by a state agency to find the water to support it at some future date. But despite that glaring loophole and the future risk it poses, the basic system has largely succeeded.

In 1978, before the act was passed, there were 452,000 acres of irrigated agriculture in Maricopa County, the central Arizona County that is home to the Phoenix metro area. By 2012, that amount of farmland had been cut by more than half, to 193,000 acres.[35] Total water pumped from the state's aquifers dropped more than 40 percent between 1980 and 2010.[36] Municipal groundwater pumping grew, but a two-thirds decline in agricultural groundwater pumping more than offset the difference. By 2010, what had been an enormous groundwater pumping deficit draining Arizona's aquifers had been largely eliminated in the "Sun Corridor," the state's growing urban region stretching from Phoenix to Tucson.[37]

Cotton, once king, had by 2015 become a bit player in Arizona's economy, with the crop planted on less than a quarter of the acreage that it had been grown on during the boom years of the early 1980s when Andrus had demanded a clampdown, and now making up just 5 percent of the state's agricultural income. Overall groundwater pumping for agriculture dropped by almost half from its 1975 peak of 5 million acre-feet per year to 2.6 million acre-feet per year in 2010.[38] From 1980, when the groundwater law was passed, to 2010, Arizona's population more than doubled to 6.4 million, but its total water use declined by 24 percent. Arizona had demonstrated that those two young economists who, half a century earlier, had questioned the state's extravagant claims about its need for water, had been right. Arizona's economic growth did not require it to use more water.

Still Picking a Fight

While the Groundwater Management Act was not a perfect law, it did lead to clear reductions in water use and demonstrated that Arizona had the tools to solve its own internal water-management problems. Arizona also has played the role of collaborator in recent years, working with its neighbors in the early 2000s to come up with a modest agreement to begin curtailing water use as the water levels in the Colorado River's big reservoirs drop.

But Arizona's approach to Colorado River Basin politics still shows strong signs of the combative state that called out its "navy" to thwart construction of Parker Dam. In the summer of 2015, as the basin states and federal government grappled with falling water levels in Lake Mead and the threat of looming shortages, Arizona governor Doug Ducey warned the audience at a civic luncheon in Tempe that California and the feds were colluding to snatch away more of the Colorado River's water. Arizona officials could offer no evidence that such a scheme was actually under way, but the rhetoric was sharp. "California," Phoenix water manager Kathryn Sorensen warned darkly, "has not shared what they're doing."[39] As Tucson journalist Tony Davis wrote, "Any threat from giant California has always been a potent Arizona rallying cry."[40]

Within the network of state and water-agency representatives working on Colorado River Basin problems, there is a clear recognition that eventually some sort of "grand bargain" will be needed that finds a way to reduce everyone's water allocation. To keep the system from crashing, everyone will have to give something up. But each of the participants in that core network also understands the dilemma that follows: each must then go home and sell the deal in a domestic political environment that views the river's paper water allocations as a God-given right. Arizona's belligerence in the summer of 2015 was a stark reminder of the way domestic in-state politics stands in the way of real solutions to the basin's problems. Without a change in attitude, Arizona's belief that "water's for fightin' over" could become, rather than a myth, a self-fulfilling prophecy.

CHAPTER 6

Averting Tragedy

The lawn at Redondo Beach High School, on the coastal edge of Los Angeles County's West Basin, began dying in the early 1940s.[1] In an arid climate like Southern California's it would be easy to blame drought, but Redondo Beach's problem was deeper. The school's gardeners, drawing on the aquifer beneath the coastal city, had enough water for the grass. The hitch was its quality. As water tables beneath the valley dropped under the stress of groundwater pumping, saltwater from the Pacific Ocean had begun to infiltrate the once-pure aquifer.[2]

A few miles inland, the communities of Inglewood and Hawthorne were either oblivious to the problem or, more likely, they willfully ignored it. Farther from the coast, they figured they had plenty of groundwater to meet their needs and (rightly or wrongly) had little fear of saltwater intrusion. They seemed not to care that their pumping might affect their Redondo Beach neighbors or even the entire region. Instead, they continued to pump, lowering the regional water table, depleting the long-term supply, and pulling the saltwater inexorably inland at the risk of contaminating the entire basin's supply.

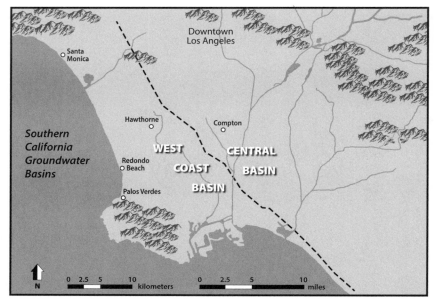

West Basin.

Like competing families sharing a fishing ground, the communities were trapped in a prisoner's dilemma: if Redondo Beach backed off on its own groundwater pumping in order to preserve the resource, that would just leave more water for the others.

Los Angeles has always been a water-challenged place. Like most cities, it grew up astride a river. The Los Angeles River (as well as the San Gabriel and the Santa Ana Rivers, also threading their way through what is now the greater Los Angeles metropolitan area) would never be mistaken for the Thames or the Hudson. But you build a city with the rivers you have, and Southern California has never lacked for ambition, water or not. At 12 inches (30 cm) of rain per year in its valleys,[3] the region falls just short of qualifying as a "desert," yet by the mid-1800s, irrigated vineyards flanked the arid region's rivers. The arrival of railroads and the evolution of refrigerated train cars made irrigated agriculture one of the region's early economic engines, and by the 1880s farmers in the West Basin were pumping groundwater to make up for shortfalls in surface-water irrigation.[4] "The story of the growth in this region," US Geological Survey scientist Walter C. Mendenhall wrote in 1905, "becomes a story of the utilization and application of its available waters."[5]

Thus it was that the LA basin, one of the first regions in the western United States to make widespread use of groundwater, also became one of the first regions forced to cope with what occurs when groundwater begins to run out.

By the 1940s, most of the West Basin water pumpers suspected there was a problem. They could see it when the depth to the water table in their wells dropped, and when the water grew salty like Redondo Beach's as the Pacific Ocean crept silently in to fill the hole left by their overuse. But while each pumper knew about their own particular situation, collectively there was an information vacuum about the condition of the basin as a whole, and how the individual pumpers' problems might be connected.

Pumping records were all private, and jealously guarded. "Individuals in one agency viewed individuals in other agencies as competitors," wrote political scientist Elinor Ostrom, whose pioneering study of the West Basin changed our understanding of the "tragedy of the commons."[6] "Water producers engaged in a quiet, competitive race with each other," Ostrom wrote in her 1964 doctoral dissertation.

Few recognized at the time what is clear today: if West Basin pumpers had continued down that path, depleted aquifers and saltwater intrusion would soon have rendered the groundwater basin useless. But even those who saw the problem also knew that if any one of them voluntarily reduced their withdrawals while the others kept on pumping, all would be for naught.

The West Basin's water dilemma was really two separate, if closely related, problems, each of which required its own institutional solution.

The first was the need to rein in groundwater pumping. If everyone kept on sucking up water as they had been, the region's water supply would fail. The second was a need for some sort of imported water supply to replace the groundwater.

Each proposed solution depended on the other to work. Without imported water, any effort to drastically reduce groundwater pumping would leave the region without a water supply. But imported water—tapping into the region's newly arrived supply of water from the Colorado River Aqueduct—was expensive. Without an effective groundwater-management scheme, communities had an enormous incentive to just keep pumping cheaper groundwater while their neighbors shouldered the costs of the imported-water fix.

In hindsight, the need for these paired solutions seems obvious; at the time, it was anything but. Redondo Beach and its coastal neighbors pleaded that collective action was needed, but Inglewood and Hawthorne wouldn't budge. It looked like the start of a western water war that could leave the West Basin an early casualty.

The Tragedy of the Commons

Ecologist Garrett Hardin in 1968 famously described the collective-action dilemma posed by situations like the West Basin pumping race, dubbing it the "tragedy of the commons." The commons is a resource to which many different users have access and cannot be excluded. A pasture on which anyone can graze their stock is the classic example, and groundwater basins from which lots of people can pump water share many of the same characteristics. By "tragedy," Hardin meant not simply a bad outcome but, quoting the philosopher Alfred North Whitehead, "the solemnity of the remorseless working of things." His formulation suggested a sort of inevitability as self-interested consumers trashed their world. A sheepherder grazing on common land has a personal incentive to add one more sheep, Hardin argued, because the benefit of that one sheep accrues to the herder. The overall harm to an increasingly damaged commons, meanwhile, is borne by all.

Such a situation need not be hopeless. Hardin suggested two possible solutions. The first is to declare the commons to be private property, and let the land's owner sort out, for a price, who is entitled to graze how many sheep. That provides an incentive to maximize the number of sheep while also minimizing damage so the pasture can support animals in the following year. The second option, Hardin argued, is an overarching government intervention to limit the number of sheep.[7]

The idea that humans would inevitably overuse a "common-pool resource" if no one owned or regulated it had been kicking around for years. But Hardin's argument and the power of his rhetoric elevated "the tragedy of the commons" into a touchstone of late twentieth-century political and policy discourse. At a time of increasing public concern about the devastation of our environment, it resonated.

With little evident opportunity or desire to turn the groundwater commons into private property, intervention by some state or federal authority to prevent over-pumping was generally seen as the only option. But starting in the 1960s with her study of the problems of the West Basin, Ostrom (who in 2009 was awarded the Nobel Prize in Economics for the answers she found) suggested a third path. She wasn't working backward from theory, telling people what they ought to do. Her starting point was empirical: What do people actually do when confronted by the sort of problem Hardin described?

The arc of Ostrom's life and work is unique. When as a political scientist she won the 2009 economics Nobel, Paul Krugman and Steven Levitt—two of the United States' best-known economists— admitted they had never heard of her.[8] This is in part because of the way her work, while nominally found in academia's political science silo, spanned disciplines. Political science rightly claimed her, economists honored her with a Nobel, and much of her research was grounded in the labor of a generation of anthropologists working around the world.

Traditional histories of California water generally treat what happened in the West Basin and the adjoining groundwater basins with a wave of the hand, as if what the communities did was straightforward—they were over-pumping, so they all got together and came up with a plan to pump less groundwater and bring in Colorado River water to replace it.[9] Ostrom's genius was in not taking that for granted, in realizing that the central question was *how* they came together to do that. Questions that to others seemed trivial to Ostrom were not. What she showed is that creating institutions capable of collectively managing the West Basin's groundwater was not as simple as slapping down a new government agency and then flipping a switch to turn it on.

Ostrom's research began as a seminar assignment while she was a graduate student at the University of California, Los Angeles. Each student was told to pick a groundwater basin and figure out how its community dealt with rising population and dwindling groundwater. It was a thorny problem, given that the boundaries of the basins and the government agencies charged with managing them did not match up.[10]

For the fledgling political scientist, the West Basin was the perfect laboratory. "It was great," Ostrom wrote years later, "that I could drive half an hour and be in 'my research site.'" Young and, by her own admission, naive about the problem she was trying to solve, Ostrom later acknowledged that "tragedy of the commons" wasn't even in her vocabulary. "I had not realized I was studying a common-pool resource problem, nor that it was widely considered insoluble," she said.[11]

It took many years, and the analyses of many more West Basin–like communities wrestling with similar problems, for Ostrom to identify a set of characteristics that successful efforts have in common. Any such governance system, she concluded, must

- define the boundaries of the area where the resource will be managed;
- determine who gets to extract the resource, and when, and how much;
- establish who pays to maintain the health of the resource so that extraction can continue into the future;
- create a process for monitoring the resource and how it is used, enforcing restrictions, and resolving conflicts;
- determine how problems across larger scales—between the resource unit itself and the larger environment in which it exists—will be resolved;
- create a framework for the evolution of rules over time as understanding of the resource and the demands being placed on it change.[12]

An Informal Start

One of Ostrom's most striking findings was the importance of informality as a tool in common-pool resource management. Ultimately, you often need to develop formal governmental and legal institutions, but at the beginning, non-binding communication—just finding a way to get together and chat—can play a critical role. Ostrom called it "cheap talk." In lab experiments, Ostrom and others found that "cheap talk" made people more likely to cooperate successfully. In the field, they found again and again, it worked.

In coastal Southern California, the West Basin Water Association provided the "cheap talk" scaffolding upon which the institutions needed to solve the groundwater problem could be built. In the 1940s, basin water management was in turmoil. Inglewood mayor E. S. Dixon was adamant: his city had plenty of groundwater, and he saw no need to join with Inglewood's neighbors in the creation of a municipal water district to solve the region's water problems. "I have consulted with geologists and other water authorities," he wrote the local newspaper in January 1947, "and all of the data convinces me that there is sufficient underground water available to the city of Inglewood to supply all of its present and future needs."[13] His neighbors in Hawthorne felt the same way. They saw only cost and risk in collaborating with their neighbors to the west.

Others had tried to convince Inglewood and other cities on the inland side of the basin that they, too, were at risk. "There is no known barrier in the West Basin to prevent saltwater encroachment," F. N. Van Norman, one of the leaders of the community fact-finding committee formed to study the problem, told the *Los Angeles Times* during the summer of 1946. "This nullifies the theory that the ultimate water problem of one West Basin city is different than any other in the area."[14]

Inglewood and Hawthorne weren't buying. The cost of joining a regional municipal water district would be high for the inland communities, Dixon said during a speech before the Southwest Inglewood Improvement Association, while the benefits of avoiding saltwater intrusion were relatively small.[15]

Inglewood and Hawthorne were gambling that the other communities would shoulder the costs of bringing in the Colorado River water and hence would reduce their pumping, leaving those who didn't join in the collective with an abundant supply of cheaper groundwater. They also were gambling that, if legal push came to shove, their water rights would win out in a court fight. That might leave the losers with no water, but Inglewood and Hawthorne didn't seem to care.

In 1945, that "I've got mine and the heck with everyone else" attitude was widespread when a group of forward-thinking individuals representing some of the most at-risk water agencies came together to form a group calling itself the West Basin Water Association. At the time, Ostrom wrote in her doctoral dissertation, West Basin water management (if you could call it that) amounted to a "competitive race among the large number of poorly informed and, in some cases, deceptively competitive water producers."[16] Initially twenty organizations joined, including the coastal cities at greatest risk from saltwater intrusion and some of the private water companies. The region's oil companies, with big economic stakes in the health of the aquifer, were active as well.[17] Ostrom chose an interesting phrase to describe them: "public entrepreneurs."

Looking back at the Water Association, it is easy to see how someone like Inglewood's Dixon would see no need to play along. The organization's structure was informal and it had no legal powers. But perhaps precisely because it was so non-threatening, it did something that in retrospect appears critical: it created a low-cost forum. Membership rules allowed essentially any agency or private entity with an interest in the basin's water management to join, and an inclusive process allowed people who weren't members to participate.[18] For community discussions about the basin's water, the West Basin Water Association created a very big tent.

This sounds weak, but Ostrom's later work on common-pool resource-management regimes found repeated examples of such a "start small" approach. Resource users begin talking about their decisions, in the process developing trust and shared understanding of the resources involved and one another's needs in using them. In short, they got to know one another.

Creating a Shared Understanding

To solve a common-pool resource problem, you first need a shared understanding of what the resource is. This sounds simple, but in the case of the West Basin, it was not.

With individual pumpers keeping their well data to themselves, the Water Association's first order of business was a fresh technical analysis of how much groundwater the basin really needed, and how much it had. The association hired an independent expert, a respected consulting engineer named Harold Conklin, to update Mendenhall's early work on the basin. The findings on the volume of overdraft, the risk of saltwater intrusion, and the cost of fixing the problem, were grim. But as is often the case in such disputes, many of the parties either contested the data or remained oblivious to the risk.[19]

There's a tendency in contentious situations like the West Basin's water battles to think science can solve the problem—that sufficient evidence can be gathered to persuade holdouts, that you can win the policy argument by winning the argument over science. Political scientists call this "scientization."[20] It doesn't work. Instead, as the reaction to the new West Basin study demonstrated, the science merely becomes a fresh focus of argument. The old arguments between Redondo Beach and its neighbors about what steps to take merely shifted to an argument over the evidence.[21]

But the West Basin Water Association also shows how "cheap talk" can create a forum for overcoming this kind of a problem. For the first time, pumpers were publicly sharing information about groundwater conditions in the basin. The group's technical files and the minutes to its meetings were open. It published a newsletter with the motto "let there be no surprises, either pleasant or unpleasant."[22]

Harold Conklin's study laid out some promising approaches to dealing with the basin's problems, including one that appeared particularly attractive—connecting the region's water users to the network being built by the Metropolitan Water District of Southern California (known as "Met") to bring Colorado River water across the desert to the coastal plain. At the time, the West Basin's communities were outside Met's service territory, so they would have to convince the agency to annex them into its boundaries to get the water. But Met's management was strongly opposed to piecemeal annexation, and for good reason. The interconnected nature of the groundwater basins made freeloading a serious risk. Met water was more expensive than groundwater. If some communities connected to Met's system and reduced their groundwater pumping, that would merely leave more water for those who did not

join. Met customers would simply be subsidizing the continued depletion of the region's aquifers.

The only way for West Basin communities to join Met was to first
join among themselves, creating a regional water agency to act as a middleman. On a rainy January day in 1947, West Basin voters went to the
polls to pass judgment on the formation of a new regional water district
to bring the area's cities under one umbrella, assessing to all of them the
costs of connecting to the Metropolitan system. The measure failed.

The final tally reflected the geographic split plaguing the basin.
Coastal residents, sitting atop increasingly salty groundwater, supported
paying more to hook up to Met by a margin of as much as five to one.
Inland residents of Inglewood and Hawthorne, feeling less threatened
by saltwater intrusion, balked at the cost, and voted "no" by a more
than three-to-one margin. That was enough to tip the balance, and the
attempt to bind the West Basin together to take collective action on its
water problems failed.[23]

Undaunted, the leaders of the coastal communities at greatest risk
regrouped and succeeded in a second election, creating the West Basin
Municipal Water District in the fall of 1947, without Inglewood and the
other inland communities that had balked the first time. The boundaries were less than ideal, with large swaths of the basin left out. But it was
not long before Dixon and Inglewood changed their minds, concluding
by 1949 that they had more to gain by joining their neighbors and participating in the decisions over who got how much water in the basin,
and how its aquifers could be protected.[24]

One by one, the other holdouts followed Inglewood and decided to
join their neighbors, until by the early 1960s the region had a robust
institution that covered the basin's major water users, with the necessary
authorities and institutional structure to both manage and distribute
imported Colorado River water. Half of the problem had been solved.
But Hawthorne, one of the largest inland communities, one of the big-

gest pumpers, and one of the handful of major water users who had never joined the association, remained a holdout.

Adjudication

Solving the problem of regulating California's West Basin groundwater pumping proved far more challenging than organizing to bring Colorado River water to the basin. But for self-governance to work, the West Basin needed both. The problem was holdout Hawthorne, happy to play its neighbors for suckers and continue pumping down the aquifer while everyone else tried to save it.

Legally, the basin's water users turned early to a process called "adjudication," a court action in which groundwater pumpers sue their neighbors, starting a formal process in which a court is asked to determine who is entitled to how much groundwater. In the old "water's for fightin' over" narrative, this sounds like a classic case of water wars, and it can be a high-stakes game. Winners can come away with a bounty, losers can see their wells go dry. But because of the risks and uncertainties, adjudication in Southern California groundwater basins proved a remarkably robust way of steering through the dangerous rapids of self-regulation. It was the prompt needed to get people to the table to negotiate a deal.

Notwithstanding Marc Reisner's complaint about the Colorado being the world's "most litigated" river, California water judge Leon Yankwich argued that the courts were a useful societal tool for settling differences. "It is the aim of litigation," Yankwich wrote, "to achieve social peace."[25] Sometimes the courts can be your friend. That is the way the water agencies were using the courts—not as a field of conflict, but as a tool with which they could fashion collaborative deals.

The adjudication process proceeded in parallel with the establishment of the new West Basin Municipal Water District to import water. But it moved more slowly. The court asked the state Division of Water

Resources to referee the matter, which it did by conducting a study whose findings were even grimmer than Harold Conklin's. If the court adopted the referee's findings, the basin's water users would have to all but stop pumping entirely. The stick wielded at this point by the adjudication process was large, creating significant appetite for the carrot of a negotiated alternative that might still allow some groundwater pumping as the region shifted to the use of imported Colorado River water, worked on the construction of a saltwater intrusion barrier, and began learning to recycle its wastewater to make up at least some of the difference.

By the early 1950s, with Met water beginning to flow, basin water users took a critical step—one that demonstrated a new level of trust. At the time, they were unable to agree on exactly how to share the pain of reduced pumping. The cuts required seemed simply too deep for many of the users. But realizing that something nevertheless needed to be done, they agreed to an interim deal in which water users would voluntarily reduce the amount of water they took from the aquifer. It was precisely the thing Hardin's "tragedy of the commons" argued would *not* happen, but after a decade of collective work, some of the users were ready to make reductions, even without binding commitments from all the others.

While negotiations continued, water users agreed to cut back to the levels they had been pumping back in 1949. It was a modest first step, but it worked. Despite the lack of a binding deal, the voluntary reductions reduced the pressure on the aquifer and the water table began rising in many areas of the basin. But it was a high-risk move, because it made free riding that much easier for those who chose not to participate in the new regime.

In that regard, Hawthorne continued to pose a problem, illustrating the shortcomings of the basin's two-pronged approach. By the early 1950s, Hawthorne had joined its neighbors by voting to annex itself into the territory of the West Basin Municipal Water District. That gave it access to the Metropolitan Water District's imported Colorado River water, suggesting that the collective management regime might finally include all the right water users. But Met's concerns about free riding were well founded. While the other users shifted to more expensive Met water in order to preserve the aquifer, Hawthorne continued to pump cheaper groundwater, playing its neighbors for suckers. The aquifer rose, but a chunk of the precious water being saved merely flowed down beneath the ground into a trough, created by Hawthorne's willful cheating, that was thirty to forty feet deeper than that of its neighbors. By using the cheaper water, Hawthorne saved $100,000 compared to what it would have cost to use Met's Colorado River supply.[26]

Hawthorne continued to sit out the negotiations and complain, but in the end this didn't matter. The other parties negotiated a compromise that represented a middle ground. More pumping was allowed than would have been permitted under the original state referee's report, but limitations were imposed on each pumper that, with the addition of collective action to build a saltwater barrier, proved sufficient to stabilize the aquifer. The parties filed the settlement with the court, which had the power to impose the resulting restrictions even on those, like Hawthorne, who had not volunteered to be a part of the deal.

Hawthorne squealed, appealed, lost, and was eventually forced by the court to join its neighbors in jointly managing the West Basin's water supply. The unpredictable threat posed by litigation, and the power of the courts, created a framework for the region's water users to work out a deal.

Enforcement and the Future

The West Basin finally had a deal, but how could they be sure that everyone would stick to the terms? Ostrom's work shows that once an agreement is in place, institutional plumbing is remarkably durable. In a study that she and her colleagues conducted of governance institutions in the West Basin and other areas in Southern California, Ostrom found sturdy networks of trust, reciprocity, and shared understanding of the resource. With reliable public monitoring, compliance was high.[27] The parties selected and funded a "watermaster" to monitor basin pumping. The data was audited, trusted, and public. But, crucially, the watermaster was not an enforcement cop. "We want to stay as neutral as possible in order to gain as much voluntary cooperation as possible," watermaster John Johams told Ostrom.[28]

Rather than the watermaster playing cop, the systems end up being self-policing. Pumpers have the option of taking a cheating neighbor to court, but Ostrom and her colleagues found that this was almost never necessary. She tells the story of the tiny Moneta Mutual Water Company, which early in the life of the agreement pumped more than its share. The watermaster, in addition to including Moneta's pumping numbers in the tables of its annual report, included a several-page-long discussion of the problem. Soon after its over-pumping was made public in this fashion, Moneta stopped cheating. Even recalcitrant Hawthorne fell into line—no enforcement required.[29]

Expanding Policy Options

In addition to regulating groundwater pumping and importing Colorado River water, the West Basin's governance system created other new policy options. This is one critical difference between cooperation and externally imposed solutions. Once communities develop the social capital—the interpersonal connections, shared understandings, and collaborative institutions—the door is open to far-more-flexible problem solving going forward.

One of the West Basin's most important new tools was a salinity-intrusion barrier, a system built along the coast to hold saltwater back. Experiments in the early 1950s showed that pumping freshwater down injection wells along the coast could create a "mound" of groundwater that held the saltwater at bay, acting like a dam to help keep it from pushing into the aquifer and contaminating wells in places like Redondo Beach. This would allow more groundwater pumping inland than would otherwise be possible, not entirely eliminating the need for communities like Redondo Beach and Hawthorne to cut back, but softening the blow. Again, the creation of such a system involved collective-action dilemmas: free riders behind the barrier would benefit whether they paid or not. And where would the water come from?

And again, the informal West Basin Water Association provided the forum for lengthy discussions on how to build an institution to do the job and what its boundaries, legal authorities, funding, and responsibilities might be. Once again, as with the West Basin Municipal Water District, the solution came to a vote of the region's residents about creating yet another overlapping joint water-management district, the Water Replenishment District of Southern California. The boundaries were different this time. West Basin water users realized that they would benefit by collaborating with their neighbors in Central Basin, the adjacent water basin to their east. But the goal was the same: create a broad institutional umbrella to accomplish collective water-management goals that no community could undertake on its own.

Again, Hawthorne mayor James Q. Wedworth balked, saying his community was already paying too much to solve regional water problems. He blamed the problem on a system that had deprived Hawthorne residents of water that should have rightfully been theirs. "If each city had been given fair and equitable water rights, the whole thing would not have come to a vote," Wedworth said.[30] But voters across the basin disagreed, approving the formation of the new district to provide imported Colorado River water as a source for the salinity barrier, and more generally to put surplus imported water back underground for future use.

Like the groundwater adjudication and West Basin Municipal Water

District, the institutions have proven remarkably robust. Over the past fifty years, the region has taken active steps to replenish its aquifers, creating a healthy groundwater buffer that communities were able to draw on during the worst droughts of the late twentieth and early twenty-first centuries.[31]

CHAPTER 7

Turning Off LA's Tap

WHEN US SECRETARY OF THE Interior Gale Norton slashed Southern California's deliveries of Colorado River water on January 1, 2003, it looked like Colorado River water *was* for fighting over. Supplies to the Los Angeles and San Diego metro areas were cut nearly in half, literally overnight. The Imperial Irrigation District, the agricultural empire that is the Colorado River Basin's largest water user, took a hit too, as the federal government applied pressure to curtail farmers' alleged overuse.[1] For years, California had been taking more than its share, something the federal government and the other states could no longer tolerate. "As secretary and river master," Norton said, "I must enforce the law of the river."[2]

California politicians had warned of the risk. "California cannot afford the immediate reduction by that amount of water," Southern California Congresswoman Grace Napolitano had said two years earlier, as the conflict was building, adding what sounded like a threat: "Our economy reaches out to the neighboring states so that if we suffer, so do the rest of the states around us."[3]

This looked like the great Colorado River water war that everyone in the basin had feared. But the events surrounding Norton's decision to cut California's access to surplus water, both the process that led up to it and the events that followed, show how wrong the myth of conflict over water is. Rather than a hostile takeover, the federal government's action represented the end result of a long, hard negotiation with the Colorado River Basin states to develop rules for sharing as water use in the Lower Basin kept growing.

Foresight

Delphus Carpenter saw this problem coming nearly a century ago. In the early 1900s, there was plenty of water in the Colorado River for the farms and cities of the sparsely populated basin to use as much as they wanted. But Carpenter, the Colorado water lawyer who was the architect of the Colorado River Compact, knew that this would not always be the case, and getting the water-sharing rules right from the beginning would be critical once demand began to overtake supply.

Carpenter was wary of the voracious water needs of rapidly growing California. In 1922, its population was greater than the other six Colorado River Basin states combined. Carpenter feared once California got a big share of the river, it would never give it up. This was a legal issue. One of the central tenets of western water law is the "doctrine of prior appropriation," the idea that the first people to put water to use have first priority when supplies become scarce. You couldn't build a new farm upstream that dried up a river for an older farm downstream. The doctrine was primarily applied within a state's boundaries, but court decisions in the early twentieth century left Carpenter afraid the rules might be applied between states as well, as decisions were made about how to divvy up water crossing state boundaries.[4]

With that in mind, Carpenter steered the 1922 Colorado River Compact toward an agreement that would hold the line against California while preserving water for later use by other states. The Compact Commission divided the river basin in half, drawing a line at Lee's Ferry, at the upper end of the Grand Canyon.[5] The states above that—Wyoming, Utah, Colorado, and New Mexico—would get half the water. The states below Lee's Ferry—Arizona, Nevada, and California—would get the other half. The upper states were growing far more slowly, but the scheme was intended to ensure that, when they finally needed it, there would still be some water left.

The agreement contained two provisions that would haunt water managers by the 1990s. The compact specified amounts of water rather than percentages—the Upper Basin would get 7.5 million acre-feet of water per year, and the Lower Basin would get 7.5 million acre-feet. The pact's authors failed to write clear rules about what would happen if the river didn't have that much water in it. Second, they built in a mechanism that ensured that unused water in the system could benefit downstream states—particularly California.

For nearly a century, the other states weren't using their full share, allowing extra water to slosh in California's direction. And just as Carpenter had feared, California had come to depend on it. By the late 1980s, California, which had an official allocation of 4.4 million acre-feet per year, routinely used more than 5 million, thanks to the other states' unused allotment. The bonus water went to the Metropolitan Water District of Southern California ("Met"), which served the growing cities on the Los Angeles–San Diego coastal plain. The other states, especially those of the Upper Colorado River Basin, were concerned "that California's growing dependence on surplus water would one day ripen into a legal entitlement," said Colorado water lawyer Jim Lochhead.[6]

Simon Rifkind, retained by the US Supreme Court to try to untangle the Colorado River mess, recognized as early as 1960 that the Metropolitan Water District had come to depend on such a surplus, but said there was no reason to worry because the surplus would be around for a very long time. "I am morally certain that neither in my lifetime, nor in your lifetime, nor the lifetime of your children and great-grandchildren, will there be an inadequate supply of water for the Metropolitan Project," Rifkind told lawyers during a 1960 hearing. "I am morally certain, as certain as I am of the multiplication table, that not within the span of the ages indicated there will be any diminution either in the present uses of the Metropolitan Aqueduct or its contemplated expansion."[7]

By the early 1990s, it was becoming clear Rifkind's multiplication table had come up short.

In response, the states were pushing the Bureau of Reclamation to do what they believed was the agency's job: forcing California to begin living within its means. "Appropriate enforcement," Utah's Larry Anderson said, "is critical to protecting our allocations under the Law of the River."[8]

"A New Era of Limits"

The easy comfort provided by surplus on the river began to slip away in 1990, when Southern California needed its usual dose of extra water, above and beyond its minimum entitlement, to tide it over. But the other basin states had had enough.

1990 was the driest year in nearly two decades in coastal Southern California, an exclamation point on a drought that had lingered, with minor breaks, for most of the 1980s.[9] Southern California usually used the Colorado River's distant watersheds as a hedge against local drought, but in 1990 the Colorado River's ability to play that role was shrinking. The first half of the 1980s had been the wettest years the Colorado River Basin had seen, leaving the reservoirs full. But beginning in 1988, nature turned off the tap, and the three years ending in 1990 were the driest to date in the Colorado River Basin.[10]

In response to local drought conditions, California government agencies took steps that would once have been unthinkable. Santa Monica developers had to pay for low-flow toilets in existing homes to conserve enough water to supply any new homes they built. Elsewhere in the basin, water districts refused entirely to hook up new customers, and the Metropolitan Water District, once known for its expansionist tendencies, stopped annexing new territory.[11] And California again turned to the Colorado River's federal managers, looking for an allocation of surplus water.

The basic dilemma will be familiar to anyone with a bank account: how much to spend now and how much to save for later? But the consequences were profound. The reservoirs in the basin serve as a massive water savings account. The rules for how much water to move, where, when, and for whom, were byzantine, but the underlying questions were simple: how much should be sent downstream for LA's lawns, swimming pools, and toilets now, and how much should be saved as a hedge against prolonged drought? Any answer to that question involved trade-offs, with someone benefiting by being able to use the water now and someone suffering the risk if the reservoirs were empty if and when the dry days arrived.

A "use it now" approach would benefit California, providing the surplus needed to keep the Colorado River Aqueduct to Los Angeles and San Diego running full. But the other states argued that this would increase the basin's vulnerability to the risks posed by a future run of dry years like those the basin had seen from the 1930s to the 1950s. Such a multiyear drought would leave the reservoirs unable to deliver the minimum legal allotment to all users. If that happened, Arizona and Nevada would be the most vulnerable, because their shares would be the first to be cut in a true shortage. California would benefit from a surplus declaration while the risk would be borne by Arizona, Nevada, and the other upstream states.

The other six Colorado River Basin states were unwilling to accept the resulting risk, and vehemently disagreed with California's request for a surplus declaration. In response, the Bureau of Reclamation set down an important marker for how the "drain or save" argument was to be resolved. In 1990, the agency's staff concluded they could find "no clear basis in the existing legal and institutional framework of the Colorado River" to give California surplus water "without the consensus of all seven Basin States."[12] All for one, and one for all. California could only get the extra water if the other states agreed.

It was a critical move by the bureau that would shape the debates that followed, and remains as an underlying principle for Colorado River management today. The bureau was wagging its finger at the states and saying, in essence, "It's up to you to come up with a way to solve this. Don't expect the federal government to take sides in your disputes over how to allocate the river's water."

Everyone understood the risks of going to court. Water managers hate uncertainty. Litigation is uncertainty writ large, and the *Arizona v. California* battle over Colorado River water allocation had created a decade of high-risk uncertainty. Where the give-and-take of negotiated solutions can add flexibility and adaptive capacity, litigation is suited to settling narrow questions or, as in West Basin, questions on which the parties all agree and the courts are merely a tool to implement a collaborative solution. As a tool to settle a conflict, litigation tends to constrain future water operations. Lesson learned.

The basic framework of a deal was clear from the beginning, but it took more than a decade to work out the details. Nature helped out, with enough wet years to help refill the reservoirs, reducing the problem's urgency, and the negotiations dragged on.

Colorado governor Roy Romer sketched out the rough terms in an August 1991 letter to California governor Pete Wilson. Romer recognized that California had a real and immediate problem: hammered by drought, California needed its surplus water in the short term to serve the parched cities of Southern California. But Romer's letter also recognized that, in the long run, the rest of the basin states needed some assurance that California would eventually learn to live within its means. California had to agree to a concrete and enforceable plan that would bring its Colorado River water use from more than 5 million acre-feet to its allotted 4.4 million acre-feet per year. The deal was to be quantified in very specific rules governing how the secretary of interior was to declare the "surplus" each year that would give California its extra water. No longer would basin water managers face the uncertainty they faced in the summer of 1990 over how much water would be available to each.

Essentially, Romer offered California what came to be called a "soft landing." "We want to help Southern California with its drought situation. But it also is in our self-interest to get California to live within its entitlements in the river," Romer said.[13]

Bureau of Reclamation officials signaled what the discussion was really all about in a presentation to one of the early meetings entitled "Managing the Colorado River in the Lower Basin in an Era of Limits." It was a recognition that Rifkind had been wrong, and that the surpluses had run out.[14]

California Struggles to Comply

The approach embodied a core principle of Law of the River. The Colorado River Compact, the Upper Basin Compact, and statutes approved by Congress set each state's overall water entitlement. It is then up to each state to figure out how to divide the water among its users. California's case was complicated by the federal government's role in administering contracts for the allocation and distribution of that state's Colorado River water. But as a practical matter, no one could come in from the outside and tell California how to get to 4.4 million acre-feet.

This is the most difficult part of making Colorado River water management work. Any agreement to deal with the basin's overall water problems inevitably must be implemented at the local level, one irrigation district, city council, and water user at a time. That is because people at the most local level are the ones actually using the water: a farmer deciding which crop to plant this year, a resident turning on the tap. State representatives may negotiate a deal that works at the basin scale, but if they can't then make it work back home, the whole thing falls apart.

At the basin scale, state leaders had agreed that California would have to live within its 4.4 million acre-feet per year allocation. In order for the other states and the federal government to give California time to get there—the "soft landing"—California had to come up with an acceptable plan to do that, with binding commitments and enforceable penalties if they didn't. They called the unspecified solution "The 4.4 Plan" or, more elaborately, the "Quantification Settlement Agreement," and the task of building a workable proposal looked intractable.

California's difficulties embodied the tension between the farm country that the Colorado River's developers originally imagined and the urban nation we have become. As the state's water managers tried to figure out how to live within its means, Californians had to contend with longstanding water allocations that gave most of the river's water to desert farms. America plumbed the Colorado River to turn desert into farmland, both out of a philosophical desire to embody the Jeffersonian ideal of the yeoman farmer, and a practical need to feed itself. A system designed to spread water across vast acres of farmland was having a hard time making the transition to the post–World War II era of growing cities. Repurposing the physical plumbing was a manageable problem, but the institutional plumbing needed to reallocate the water was another matter. The rules were all wrong, and the politics made it difficult to change them.

In 1931, the Southern California agencies that used the most Colorado River—the Palo Verde Irrigation District, the Imperial Irrigation

District, the Coachella Valley Water District, the Metropolitan Water District of Southern California, the City of Los Angeles, the City of San Diego, and the County of San Diego—signed an agreement that gave farmers first dibs on the bulk of the water—87.5 percent—released from Hoover Dam.[15] The Metropolitan Water District was entitled to the other 12.5 percent. It was a recognition of the doctrine of prior appropriation: the farm districts had a legitimate "first in time, first in right" claim. It was also consistent with the reality of water in that day and age. Whether measured by volume or cultural importance, water's primary use was on the farm.

But the deal included a safety valve that, by the 1990s, was coming back to haunt Southern California. Beyond the initial allocation of 4.4 million acre-feet, the so-called Seven-Party Agreement said that if there was a surplus, cities could use it. That surplus had been keeping Metropolitan happy all these years, enabling Los Angeles and the other cities of coastal Southern California to grow. Met routinely took double its basic allotment. But by 1990, as urban demands grew and pressure rose from the rest of the Colorado River–using states, the scheme's flaws were becoming increasingly apparent. It was not just California water users in general who were vulnerable if the state was cut back to 4.4 million acre-feet. It was one class of water user in particular—the cities. In hindsight, it is easy to say simply that Southern California should not have allowed so much growth, but by 1990, such hindsight did not help much in dealing with the facts on the ground. Changing the allocation would require the cooperation of the big farm districts that controlled the bulk of the water, and their political supporters. And therein lay the problem.

Since at least the 1980s, Californians had been fighting over whether the Imperial Irrigation District was wasting water.[16] The battles had ended, unproductively, in court. So the agencies turned instead to voluntary deals under which the cities would pay for agricultural water-efficiency improvements in return for a share of the saved water. On paper, this should have led to equitable arrangements. Economists love this sort of simple "efficiency" argument: let the markets work and the problem will take care of itself. But as we have seen, simply taking farmland out of production eats away an agricultural community, creating significant political opposition. And the law tends to give those farm communities the power to block change if they don't like the terms of the deal.[17]

Some low-hanging fruit was obvious. Earlier agreements had lined previously unlined irrigation canals that carried water from the Colorado River through the desert to the farmlands of the Imperial and Coachella valleys. This required no farmland to be taken out of production. But that only saved 2 percent of California's Colorado River water use. A broader agreement, which would require more farm water to be moved to city use, proved elusive. Within California, things were a mess.

The two biggest farm districts, the Imperial Irrigation District and the Coachella Valley Water District, couldn't agree on how to divide up their shares of the agricultural water. The two biggest municipal agencies, the Metropolitan Water District of Southern California and the San Diego County Water Authority, couldn't agree on terms for the use of Met's canals to move water from desert farms to urban users in the San Diego area. They couldn't settle on a plan to manage environmental impacts from the deal. And no one could agree on how to save enough water overall to make the deal work.[18]

By 1997, a frustrated interior secretary Bruce Babbitt issued a warning: if Southern California water agencies didn't get their act together and come to a mutually acceptable deal, he would be forced to step in and slash water allocations for them.[19]

Prodded by the threat, California's water agencies finally worked out a framework for a deal. San Diego would fund a slowly expanding conservation effort in the Imperial Irrigation District, with the saved water moving to the cities. San Diego and Met reached a deal on the use of Met's canals to move the saved water. The Coachella and Imperial water districts settled their dispute over who was entitled to how much of the desert's agricultural water supplies.

In the basin, there was a bureaucratic sigh of relief. It seemed that the water war had been averted. In 2001, Babbitt, in the last public event of his eight-year tenure, stood at a San Diego ceremony beneath a banner that said "Peace on the River." The deal announced that day formalized the "interim surplus guidelines," rules striking a balance between releasing enough water from the Colorado River's reservoirs to give California a "soft landing" and holding enough water back in savings to protect the other six basin states from the risk of future shortage. But there was a caveat. California's deal, while close, was not quite done, so the basin-wide agreement included a firm commitment from California to finalize its 4.4 Plan by December 31, 2002, along with a threat: if California failed to come to terms with its own problems, there would be no soft landing.

Again, the final California negotiations lagged, and the other states worried. Would the federal government have the nerve to wield the enforcement hammer come January 1, 2003? With the clock ticking, California's negotiators came to what they thought was a deal on the terms of the Quantification Settlement Agreement with less then two months to spare. Bob Hertzberg, one of the senior statesmen of California politics and the man brought in to broker the deal, again trotted out the rhetoric of "lasting peace on the river."[20]

But without the approval of the board of the Imperial Irrigation District, the Colorado River's largest farm-water agency, Hertzberg's deal would collapse. All eyes in the basin focused on an emotional meeting of the Imperial Irrigation District board of directors on December 9, 2002. The farm district was under enormous pressure, and there were fears that if they did not agree to a deal that included compensation for their water, it would simply be taken away.[21]

Gustave Aguirre of the United Farm Workers urged the IID board to approve the deal only if there was an added $70 million for economic transition assistance for the chronically poor communities of the Imperial Valley. Antonio Ramos of Calexico dramatically held up a milk jug with little water in it and recalled the other infamous agricultural-to-urban transfer in California that dried up basins in the eastern Sierra Nevada in order to supply water to Los Angeles: "If it goes through we will have no water and the valley will become another Owens Valley—dead."[22]

In the end, the board voted three to two against the deal.

Federal officials wasted no time in responding. Assistant Secretary of the Interior Bennett Raley wrote a letter to the California delegation saying that the failure to formally quantify how much water would be shifted from the Imperial Valley to the coastal cities meant California had failed to meet the terms required to earn the soft landing they'd been hoping for. In what is surely the most literary moment in the history of the Law of the River, Raley referenced F. Scott Fitzgerald: "The Department has no interest in a (Quantification Settlement Agreement) that does not represent a long-term Quantification of the parties' portion of California's apportionment of Colorado River water, lest in fifteen years we find ourselves as Gatsby did—'So we beat on, boats against the current, borne back ceaselessly into the past.'"[23]

Thus, on January 1, 2003, the big pumps that push water from Lake Havasu up 594 feet to Copper Basin reservoir and the start of the Colorado River Aqueduct were throttled back, and California began the difficult task of living within its 4.4 million acre-foot Colorado River allocation. Slowly, also, the cutbacks filtered down to a reduction in agricultural water diverted at Imperial Dam. California was forced to share.

Whether this looked like success or failure depended entirely on where you sat. California's attempt to negotiate a "soft landing" had failed. Efforts by the other six states to ensure that California lived within its 4.4 million acre-feet per year allocation succeeded.

Yet there was another, more subtle element of success that was not apparent on January 1, 2003, as the water in the Colorado River Aqueduct slowed. Over the previous decade, through what at the time had appeared to be chaotic and often unproductive arguments, the river's managers had been building an institutional framework for life in the era of limits. The contentious process had created a new approach to sharing the Colorado River Basin's water. Success was not guaranteed, but the right tools, both within California and at the scale of the entire Colorado Basin, had been identified and were ready to be put to use.

CHAPTER 8

So Cal Cuts Back

W<small>AS</small> S<small>OUTHERN</small> C<small>ALIFORNIA</small> headed into a crisis as a result of the draconian 2003 cutback in Colorado River water? Recall the words of Congresswoman Grace Napolitano, who represents a slice of suburban Los Angeles County, in the run-up to the cuts: "California cannot afford the immediate reduction by that amount of water."[1]

Napolitano's warning followed a pattern familiar in the arid western United States. Fear of water shortage is greater than reality, as communities underestimate their ability to cope when supplies run dry. When people have less water, they use less water, often with greater ease than they thought possible. What happened in Southern California in 2003 demonstrates that clearly.

In public, area water managers were blasé. "There is clearly no emergency, due to Metropolitan's foresight and planning," Southern California Metropolitan Water District (Met) board chairman Phillip Pace told his colleagues at a January 6 emergency meeting to discuss the agency's options.[2] True, Pace was putting the best face possible on a difficult situation. Behind the scenes, managers were scrambling to make ends meet.

They believed there would be enough water from the State Water Project, which brought Northern California water south across the mountains to Los Angeles and San Diego, to cover the 2003 shortfall in Colorado River water. But the plumbing issues were complex: different treatment methods were needed to balance the two sources of supply, and the necessary interconnects to get State Water Project water to communities that previously got their water from the Colorado were problematic.[3]

Nevertheless, by 2003 Met was a changed system from the old days when it simply ran big aqueducts full blast. Following drought in the late 1980s and early '90s, Southern California had made significant improvements—storage deals with Central Valley farmers, conservation initiatives, dry-year options to move water from farms to cities—leaving its water-delivery architecture far more flexible and robust, able to weather a significant shock, even if the specifics of the loss of Colorado River Aqueduct water were poorly anticipated.

The Laguna Declaration

Southern California was not always so thrifty. Like every major metro area depending on the Colorado River, it was built on imported water that was moved out of the river corridor to adjoining land that would otherwise have been dry. Albuquerque, Denver, Cheyenne, and Salt Lake City literally move the water physically out of the basin. Las Vegas, Phoenix, and Tucson are technically in the Colorado River Basin, but they have to pump the river's water back uphill to use it. Initially, like the others, Southern California cities overreached their groundwater and built increasingly long pipelines and aqueducts to augment supplies. This began with the Los Angeles Aqueduct to the Owens Valley, completed in 1913, followed by the Colorado River Aqueduct, which delivered its first water in 1939, and the State Water Project, which moved Northern California water to the Southland over the Tehachapi Mountains beginning in the early 1970s.

The political geographies of modern American cities are complex things. Los Angeles city government built the Owens Valley aqueduct, but the city limits encompassed only a fraction of the region's population. To gain access to its water, communities had to give up their sovereignty and annex into LA. Preexisting cities like Pasadena had no intention of doing that, so in 1928 they banded together and formed the Metropolitan Water District of Southern California, a regional water agency that would build the aqueduct to bring water from the Colorado River, spreading the water beyond the boundaries of a single municipality to an entire region. As the federal government was preparing for the construction of Hoover Dam, Southern California was preparing the governance structures needed to build the plumbing and use the water.

But the creation of Met, initially encompassing Los Angeles and twelve suburban neighbors, simply raised a different version of the same question: what if neighboring communities wanted to join in the project and share in the water? The boundaries of who was in and who was out were still an issue. The initial Colorado River supply was far more than needed by Met's thirteen charter members, leaving enough water for more communities to join the club. But who, and how many? In 1938, Met's Water Problems Committee drew a line: the ultimate capacity of the Colorado River Aqueduct would set the limit, and decisions about annexation and extension of supply would depend on there being an adequate supply of water.[4] Water scarcity would define the boundaries of how big Southern California's cities could grow.

Met governance has always been politically complicated. The largest and most powerful municipal water agency in the United States, Met functions as a water wholesaler, with a governing board made up of representatives of each of the municipal member agencies that pass Met water along to homes and businesses. Its decisions influenced critical growth policy as Southern California navigated its transition from an agricultural empire to a megacity.

For many of the people operating at those boundaries between water supply and urban growth, the old 1930s thinking about limits simply would not do. As the state of California eyed a massive north-to-south project to bring water from the Sacramento River all the way to Southern California, the region's postwar growth boosters wanted a piece of the action, as always in the arid West viewing water shortage as the defining constraint on economic growth. The boosters could see a time when they would grow into and eventually surpass the Colorado River allocation. They were keen to ensure that a shortage of water would not get in the way of Southern California's never-ending boom. At the boosters' side was Met, which acted as a sort of shadow regional government when it came to getting water supplies to the region and parceling the commodity out to local water agencies.

Rapid territorial expansion pushed Met's territory to the east, into valleys that remained at the time largely agricultural but were destined to become rapidly expanding suburbs. In 1952, the Met board, meeting in the coastal town of Laguna, made the policy official in what came to be known as the "Laguna Declaration," an explicit statement that Met's job was to get whatever water the booming metropolis decided it needed: "When and as additional water resources are required to meet increasing needs for domestic, industrial, and municipal water, the Metropolitan Water District of Southern California will be prepared to deliver such supplies."[5]

That "tell us what you need and we'll get it" philosophy dominated Met's management approach for some four decades, and the agency did just that. From the east, it ran the Colorado River Aqueduct full bore, an artificial river larger than the Rio Grande pumping water through mountains to the Southern California coastal plain. California kept the aqueduct full in part by soaking up surplus supplies that other Colorado

River Basin states weren't using. With the Laguna Declaration in hand as its guiding principle, the agency added the California State Water Project, the region's third great artificial river, bringing water south from Northern California by the early 1970s. Initially, Met pursued its own canal project, which would have tapped the Eel River on the state's rainy northern coast to bring water to the cities of the Southland. But eventually political reality made clear that the only way to pull off a project that big was a statewide effort that brought together the interests of farmers in the state's big Central Valley (who had their own problems with groundwater shortfalls) and the cities of the south.[6]

With 701 miles of pipelines and canals, twenty-one reservoirs, the ability to irrigate 750,000 acres of farmland and serve 25 million people,[7] the State Water Project is a staggering hydraulic achievement. Southern California's leaders were sure, with three independent sources of supply—one the state aqueduct from the north, the second the Colorado River aqueduct from the east, and the third serving the city of Los Angeles from the Owens Valley—that their water supply was secure. They were mistaken. Southern California first overdrew the available groundwater supplies, and then it overdrew the Owens Valley aqueduct. When drought began in the late 1980s and lingered into the early '90s, their confidence was shaken. "We were wrong," said Ronald Gastelum, Met's president and CEO. "The state water project did not produce. Fortunately, we had the Colorado River supplies to rely upon and we barely got through that crisis. It was a wake-up call for us."[8]

This time, the region's leadership faced up to the reality that simply adding new big aqueducts to import more water would no longer work. Met and its allies in Sacramento had already lost a bruising statewide political battle in the 1980s over construction of a new Northern California canal to help send even more water from the state's wet north to its arid south, and options to move more water from outside the region were looking increasingly impractical.

The policy shift took the form of a 1996 "Integrated Resources Plan" that turned the Laguna Declaration's idea of delivering whatever water its members said they needed on its head. Rather than bringing water in from the outside, the plan called for being shrewder with the water they already had. Recycling was a major component—cleaning up sewage and putting it to a second use rather than just dumping it in the ocean, as had long been done. The oddly named strategy of "groundwater recovery"—cleaning contaminated groundwater so that it could be used—played a role. Expanded use of aquifers to store water during wet years so that it was available during dry years provided a buffer. They also built a new Southern California reservoir, in a place called Diamond Valley, as a hedge against drought. And desalination of both ocean water and brackish inland groundwater was added to Met's toolbox.

The document, which grew out of a three-year planning effort, opened with unusual literary flair, quoting Wallace Stegner: "And more important . . . was one overmastering unity, the unity of drought." Its language marked a major conceptual shift in the relationship between Met and its member agencies. No longer would Met simply bring the water. Managing under the conditions Stegner described would require coordinated approaches across the boundaries between large agencies and small.[9]

In keeping with the Laguna Declaration's underlying philosophy, however, limiting growth was never an option for bringing long-term supply and demand into balance. Met would continue to support whatever population might inhabit the region.

The plan included a conservation push that was unprecedented in Southern California, where lawns and swimming pools seemed a God-given right. That was something that was already under way, beginning in the drought of the late 1980s, but the Integrated Resources Plan locked it in as a matter of regional water governance policy. Per capita water use had been dropping, but not fast enough to make up for

Southern California's population growth. If Met had reached the limit of available imported supplies, more steps would be required.

But the 1996 Integrated Resources Plan also included what, in hindsight, was a critical mistake. The document discusses the risk that Southern California might not, in the future, be able to depend on the surpluses it had been using to keep the Colorado River Aqueduct full. But when it came time to total up the final numbers, Met's planners concluded that their full allotment of 1.2 million acre-feet per year would be reliable at least through 2020.[10]

Just seven years later, they found out how wrong they were. But they also discovered that the water-management widgets they had built in the meantime—things like recycling, conservation, and groundwater storage—left their system robust enough to handle a problem they had not anticipated. Southern California passed a resilience test caused by a shock larger than the region's water managers had planned for.

Water Storage Deals

When Met's Colorado River supply was slashed in 2003, one of its key safeguards was a series of groundwater storage agreements it had developed over the previous decade. The basic principle was to find a place where Met could store surplus underground in wet years, to be called on in dry years. In all, this surplus amounted to more than a million acre-feet of groundwater stored outside of Met's primary service territory—an amount equivalent to nearly an entire year's worth of Colorado River supply.[11]

These agreements belie the popular image of California city dwellers and farmers in a death struggle over scarce water, reflecting instead collaborative relationships that maximize resources. Met's agreement with the Semitropic Water District located near Bakersfield in California's Central Valley is an example of how this works.

Semitropic is a mountain range away from the Colorado River Basin, but the water-swap agreement with Met signed in 1994 shows how a combination of physical and institutional plumbing can help balance out the ups and downs of water supplies in the two basins.

The managers of Semitropic had begun offering their aquifer, for a price, to store surplus water from the California State Water Project, the system of pumps and canals that delivered water from California's wet north to its arid south. Whereas for most of its life the Colorado River Aqueduct had delivered a stable supply year in and year out, the State Water Project deliveries were always more variable, depending on whether the Sierra Nevada, which feeds the system's water, had a big snowpack or a small one. Storage would help to even out those ups and downs. Serving a marginal Tulare Basin farming area that had long relied on groundwater, Semitropic connected to the State Water Project beginning in the 1960s. During wet years, Semitropic's water banking project allowed State Water Project contractors who had more water than they needed to send the extra to Semitropic. The agency would either divert the water to spreading basins to recharge its aquifer, or provide it to farmers in lieu of groundwater the farmers would otherwise have pumped.[12]

For Semitropic, the financial benefit was significant, an example of a rural-urban partnership that uses water-management income to help support agricultural communities. In 2012, for example, the agency made $14 million from its water-banking customers, the same amount of revenue that it got from the local customers actually using its water.[13]

Aquifer storage and recovery, used in a wide variety of settings, offers the same basic function as a dam: hang on to water now so that you can use it later. It offers both advantages and disadvantages over dams. On the plus side, there is no evaporative loss, you don't have to block off a free-flowing river, and you don't need to shoulder the huge up-front capital cost, usually borne by state or federal government, to make it happen.

On the minus side, there are some losses, because you can never reclaim all the water you put into the ground. In the short run, that means reservoirs are more efficient despite their evaporation, though it only takes a few years' storage to make aquifers a better deal. In addition, getting the water out requires a bit of cleverness. To work, for example, the Semitropic-Met agreement also requires a second bit of water-management cooperation and flexibility between the two water agencies when the time comes to take the water out. Semitropic and Met have two options. Semitropic has the ability to pump water out of the aquifer, moving it back through its canals to the State Water Project delivery system to be shipped on to Southern California. But the best option is simply to do an accounting swap. Semitropic farmers pump out Met's groundwater to use on their farms, and a portion of Semitropic's allocation in the year it is used is left in the aqueduct to be pumped on to Southern California.[14]

By the time the Colorado River shortage hit in 2003, Met had stashed more than 400,000 acre-feet of its surplus Northern California water in Semitropic's aquifer, along with similar stored water with three other California agricultural water agencies. In the first year of Colorado River shortfall, Met didn't need to call on its Semitropic water, getting enough to carry it through the year from a similar groundwater storage arrangement with the Coachella Valley Water District, plus other water stored within the agency's system in various places around Southern California. Over time, as drought continued, Semitropic became critical to keeping water flowing to Southern California's water users. The physical and institutional systems, operated in tandem, gave Southern California the resilience to withstand the Colorado River Aqueduct curtailment.

With the groundwater storage and other innovations developed as part of the 1996 Integrated Resources Plan as backup, Southern California weathered the 2003 cuts without significant impact on its customers. Bill Hasencamp, Met's Colorado River Resources Program Manager, likes to point out that not only did the problem of a massive cut in Met's supplies not make the front page of the *New York Times*, it did not even make the front page of the *Los Angeles Times*.[15] Far from being the sort of existential crisis that in earlier days might have triggered dire headlines, the shortfall had become a manageable problem.

CHAPTER 9

The Great Fallowing

THE 2003 DECISION TO SLASH Southern California's Colorado River allotment forced an unprecedented and often uncomfortable partnership between farmers and cities, as the region attempted water conservation and agriculture-to-urban transfers on a scale never before seen in the United States. The urban water agencies of Southern California and the rural Imperial Irrigation District had been trying for decades to jointly manage the region's Colorado River supply, but the relationship had never been an easy one.

The Southern California Metropolitan Water District (Met) represents the rich city relatives and the Imperial Irrigation District the poor country cousins. Per capita income in Met's service territory of San Diego, Orange, and Los Angeles counties is nearly double that in Imperial County. But Imperial County has always been rich in water, to which the community's leaders cling tightly.

For shear audacity in our confrontation with North America's arid West, it is hard to top the Imperial Valley in California's southeastern corner, near the bottom of the Colorado River system. Driving west on

The All-American Canal, carrying water to the Imperial Valley (© John Fleck).

Interstate 8 from Yuma toward San Diego, you pass through high desert dunes and then scraggly scrubland before descending a small hill and across an artificial river known as the East Highline Canal. The canal marks the sharp dividing line between desert and a blanket of green—a frenzy of lettuce and carrot harvest in winter, vast acres of durum wheat, kleingrass, sudan grass, and leafy green alfalfa in spring and summer—draped across the heart of what was once a desolate, nearly impassable tract of dry land.[1]

It is Colorado River water working its greatest magic, diverted at Imperial Dam on the Arizona-California border, shunted from its natural channel into a canal that by itself is larger than Arizona's largest rivers, the Salt and the Gila, or New Mexico's Rio Grande. That big artificial river, the All-American Canal, distributes water to a spiderweb of some 1,600 miles of irrigation ditches to deliver water and drains to take excess away from farmers' fields, creating one of the largest irrigation districts in the nation.

History and law give the Imperial Irrigation District—known as "IID"—title to more Colorado River water than any other region or entity—more than Los Angeles, far more than Las Vegas, more at the Imperial Irrigation District's water-usage peak than the entire state of Arizona. In 2013, the valley's farmers used that water to grow $2.3 billion worth of agricultural products, including a piece of the vegetable market that, with Imperial County's neighbors across the river in Yuma, provides nearly all of the winter greens eaten in the United States and Canada. If you eat salad in the United States or Canada in January, you are almost certainly consuming Colorado River water.

The Imperial Valley has all of that water, and is therefore able to grow all of that lettuce, wheat, and hay, because of the legal and political structures built around "prior appropriation." The valley's farmers were among the first to put the Colorado River's water to "beneficial use," and that position has carried great weight in the decisions that have followed. The water-management struggles that have resulted represent a collision of two of the great western water-management myths—that water flows uphill toward money, and that prior appropriation stands in the way of a solution to our water-allocation problems.

In both cases, the relationship between the Imperial Valley and urban Southern California show the fallacy of the myths. The coastal cities may be far richer, but even after decades of wrangling, the Imperial Valley still has most of the water. But neither have the Imperial Irrigation District's senior water rights prevented mutually beneficial deals. Urban Southern California and IID have been able to work together to move some of the agricultural district's water to the cities while still preserving the farm communities in places like El Centro and Brawley that depend on it.

The enormous size of IID's supply has made the valley's water users central players in the search for a way to reallocate Colorado River water as supplies grow scarce. Critics argue that Imperial Valley agriculture is wasteful and makes no sense,[2] but the valley's farm community has hung on, defending water rights carved out of the Colorado River a century ago, while simultaneously adjusting to an era of scarcity.

There has always been a gun-to-the-head element in the relationship between IID and the powerful Metropolitan Water District of Southern California. It is not uncommon for the district's representatives to arrive at gatherings of the Colorado River Basin network of managers and deal makers only to find that the discussion is already well under way without them. "IID's always had this 'island nation' status," said Tina Shields, an engineer who manages the district's Colorado River water.[3]

The tension arises in part because IID, more than any other big water user, operates in a local political climate that is sometimes openly hostile to the deals that the irrigation district's leadership has tried to craft. It is a common response in farm country. "Water carries a lot of emotional value to farmers," said Bart Fisher, a prominent farmer and Colorado River water user in nearby Blythe, California.[4] The Imperial Irrigation District's board is democratically elected, so community sentiment matters more to its water-management decisions than in many more technocratically managed Colorado River Basin regions.

Fighting to hang onto water, and a fear of richer, more powerful neighbors coming to take it away, are deeply ingrained in the community's culture. "Water founded this county," the *Imperial Valley Press*'s editors wrote in 2007, "and protecting our allotment will ensure its survival."[5]

The debates have been legal and political, sometimes confrontational and sometimes collaborative. Under the agreements, the Imperial Irrigation District cut water use by 20 percent in the first decade of the 2000s, and Imperial's farmers have adapted. Over the same period, farm incomes (adjusted for inflation) have risen more than 30 percent. Imperial Valley agriculture has demonstrated what adaptive capacity in the twenty-first century's "age of limits" can look like. But as climate change decreases flow in the river while population across the basin rises, the pressure is unlikely to let up.

Imperial

The valley's audacity begins with the name of the place, "Imperial," slapped onto the desert by early boosters trying to use marketing and the promise of water to turn their vision into profit.

"No other enterprise in the Southwest has in recent years made the same rapid progress in the way of home building and general agricultural development," the valley's first large-scale commercial land developer wrote in a 1903 ad in the *Los Angeles Times*. The valley, the ad proclaimed, offered "an unlimited amount of water, fertile soil, a warm climate."[6]

This has always been a tough sell. "Warm climate" is an understatement for some of North America's hottest, driest deserts. In 1903, "unlimited water" was more hype than reality, though "desert with occasional massive floods" would probably not have made as strong a marketing pitch. "Why California should feel any desire to claim the wilderness of sand and rattlesnakes lying between Vallecito Mountain and Fort Yuma, I cannot see," wrote Josephine Clifford of her 1870s trek through the harsh landscape. "Can anything be more hopelessly endless—more discouragingly boundless—than the sand-waste that lay before us?"[7]

But others saw what Clifford did not, a quirk of geography that came to dominate the trajectory of human use of the Colorado River. The "wilderness of sand and rattlesnakes" was, in fact, rich soil spread across the desert when the nearby Colorado River periodically spilled its banks. This rich soil was the crumbled remains of the rocks the river and its tributaries had carved from canyons upstream, dumped as the river slowed and built a floodplain across its delta.[8] Crucially, a large part of the valley floor was a sink that lay below the level of the Colorado River as it traveled along the edge of the mountains to the east. Gravity was all it would take, the valley's early boosters realized, to turn the desert green.

That insight—that the rich expanse of Colorado River Delta desert soil sat lower in elevation than the flow of the river itself—shaped everything that followed. This was before the age of the great pumps that now move rivers of water up out of the Colorado River's main channel to Phoenix and Los Angeles. Those human-made rivers today, driven by massive pumps upstream from Parker Dam, are each similar in size to the entire flow of the Rio Grande, the river just across the Continental Divide. But in that time before pumps were large enough and economical enough, it took gravity and audacious insight to realize that the Colorado River's waters could be moved.

The lore of the Imperial Valley credits that insight to a man named Oliver Meredith Wozencraft. A doctor from New Orleans, Wozencraft was drawn like many of his generation by the chance of riches to the goldfields of California. Traveling the southern route, Wozencraft made it as far as Yuma, Arizona, on the east bank of the Colorado River. Between him and the goldfields lay Clifford's sand-waste. Warned about the dangers of the trek across it in the late spring heat, the lure of gold was too strong, and Wozencraft and a group of fellow travelers set out.

This was very nearly the end of his life story. As Wozencraft told the tale later, he was near death, delirious with heat and dehydration when he stumbled into an old, dry river channel and had a vision: "It was then and there that I first conceived the idea of the reclamation of the desert."[9]

All that was needed, Wozencraft believed, was to divert a portion of the Colorado River's water from its natural course into that old river channel. Instead of flowing toward the sea it would bring irrigation water to the fertile desert land. Rescued when he was near death, Wozencraft devoted the rest of his life to that dream.

There is every reason to believe Wozencraft embellished the tale, that he really came to the idea of irrigating the desert later in life, when his head was clearer. Historians think it was a young geologist named William Blake, part of an early team searching for a railroad route to the Pacific, who first snapped to the idea that the valley that would come to be known as Imperial really lay below the level of the nearby river.[10] But, wherever the idea came from, there is no doubt that Wozencraft was the first to seriously pursue moving water out of the Colorado River's natural channel to irrigate the Salton Sink, and he spent the rest of his life unsuccessfully trying to turn the Colorado River's water out of its banks and toward the valley to its north. He died in 1887 in Washington, DC, at age seventy-three, thirty-eight years after he first ventured from Yuma into the desert, still waiting in frustration for the federal legislation needed to bring his scheme to fruition. "His most favorite hobby," his obituary in the *San Francisco Daily Evening Bulletin* explained, "was that of converting the desert of Southern California into a productive field by flooding it from the waters of the Colorado River."[11] He was very much a man ahead of his time, trying to convince the federal authorities that "reclaiming" a desert was a task worth undertaking.

It is impossible to replay the tape of history without the happenstance of Wozencraft's arrival to see how things might have happened differently, but it seems likely the creation of the Colorado River Basin's greatest irrigation empire would have happened anyway, even without Wozencraft laying the groundwork. The pull of gravity downhill from the Colorado River to the rich delta lands of the Imperial Valley, combined with the inexorable pressure of immigrants moving westward trying to profit by bringing the desert lands into the nation's agricultural economy, makes today's Imperial Valley agricultural empire or something like it seem inevitable. Blake's insight represented facts on the ground. Someone likely would have brought water to what remains the greatest agricultural empire in the Colorado River Basin.

But inevitable or not, the vast working landscape of the Imperial Irrigation District is now here, with people who have built their lives around it. Adapting Colorado River management in an age of scarcity compels us to take that reality into account.

Chaffey and Rockwood

It was left to a pair of entrepreneurs named George Chaffey and Charles Rockwood to carry out Wozencraft's vision, digging a canal that looped south into Mexico before joining up with the north-flowing Alamo River back into the Imperial Valley. The plan for the profit-making venture was that immigrants would settle the land and buy their water from the California Development Company, keeper of the canals.

This did not go well. Perpetually undercapitalized, the company was no match for the river. In 1905, the Colorado tore out the company's flimsy diversion gates and the river's entire flow thundered past modest irrigation works meant to contain it. The flood destroyed farms, and it took the might of the Southern Pacific Railway (at the request of the president) until 1907 to finally close the breach and put the Colorado River back into its channel, headed properly back southward to the Sea of Cortez.

There was a national debate at the time over whether private enterprise or state and federal agencies should lead the effort to bring irrigation water to the arid landscape.[12] The dramatic failure of the California Development Company helped tip the debate in the direction of federal projects, making it clear that the only entity large enough, with deep enough pockets to "tame" the Colorado River, would be the United States government.

They needed a dam, and surveyors had long known where to put it. The canyon country near present-day Las Vegas where the Colorado River sliced through a narrow gorge was perfect for the job. But the engineering undertaking required to do it was every bit as audacious as the notion of farming the desert, something only the federal government had the resources to accomplish.

The 1922 Colorado River Compact among the seven US states cleared the way for federal legislation needed to build the dam. In 1928, Congress passed the Boulder Canyon Project Act, authorizing federally funded construction of Hoover Dam, to eliminate the floods that jeopardized the lower river valleys and even out the river's flow, making steady year-round irrigation possible. And by the 1930s, a great new canal was built entirely within the United States to bring the Imperial Valley its water—the All-American Canal.

At that point, coastal Southern California's water needs were relatively modest, but over time, as the coastal metro areas grew, so did the tension between them and the desert agricultural regions of Imperial County and the valleys around it, Palo Verde to the northeast and Coachella to the west. In keeping with the principle of prior appropriation, the

The All-American Canal, larger than many of the West's rivers (© Lissa Heineman).

original water-allocation deal signed between California's water users and the federal government in 1931 gave the bulk of the water to the farm districts. A long period of stability followed, as Imperial Irrigation District farmers grew into their allocation, building out communities across the desert based on farming the desert with their Colorado River water. In 1934, before the arrival of the All-American Canal, Imperial County farmers were irrigating 245,000 acres of farmland. By the time the US Census Bureau had tallied the acreage in 1978, that had risen to 442,000 acres. But as populations grew in Los Angeles, San Diego, and the other cities and suburbs along the Southern California coast, the stage was set for conflict over reallocation of California's Colorado River supply.

Too Much Water

The modern era of legal struggle over the Imperial Irrigation District's water use began in 1980, ironically, with concerns about there being *too much* water. The complaint that the Imperial Irrigation District was using water wastefully came not from another Colorado River Basin water user, but from one of Imperial's own. John Elmore, who owned land at the edge of the Salton Sea, complained that excess runoff from wasteful farming practices in the valley were causing the sea to rise, flooding his farm.[13]

The Salton Sea has always been an anomaly. Before Rockwood and Chaffey diverted the Colorado River to Imperial Valley farmland beginning in 1901, the "Salton Sink" was a dry salt flat. When the river broke its banks, it created an inland lake in the two years it took for the Southern Pacific to heal the breach and redirect the Colorado River back to the ocean. The desert is so hot that those remnants would have quickly evaporated back to empty salt flat were it not for valley farming. But irrigation always involves infiltration and runoff, and all that water flows into the Salton Sea, replacing the evaporating water.

Thus by the 1970s, the basin's unique geography, which had made it so attractive to the first generations of farmers, was coming back to haunt water managers of Elmore's era. Because it was a closed sink, the agricultural drain water simply flowed to the "sea," accumulating there in whatever balance existed between inflow and evaporation. The balance had always posed a tricky dilemma for the irrigation district. As early as the 1950s, it instituted conservation programs to try to keep the sea's level stable, in part because the district itself had become a major landowner.[14]

Those measures, combined with drought that reduced natural inflow from regional rains, were enough to keep things in balance until the late 1970s, when a series of wet years caused the sea to begin rising again, triggering a flurry of complaints from Elmore.[15]

Elmore blamed the district's water-management practices. Canals ran full whether the water could be used or not. Farmers routinely ordered more water than they needed. That resulted in excess "tailwater" that the district made little effort to recover and divert to other farms, Elmore charged. As the sea rose, Elmore had to build dikes to protect his farmland, and install pumps to get rid of his own agricultural drain water, because his land was now below the level of the sea.[16]

Elmore's beef was about the damage to his property. But his legal strategy created an opportunity that led to a fundamental shift in the political alliances among California water managers that lingers to this day. Rather than claiming that the irrigation district's practices were harming his land, Elmore argued instead that the district was violating state law by wasting water, rather than putting it to beneficial use.[17]

This created an opportunity for the Metropolitan Water District. Desperate for water because of the failure of a scheme to bring more water to municipal Southern California from the north, Met was again looking east, toward the Colorado River. Elmore and his attorneys suggested that the Metropolitan Water District pay for the systemic improvements needed to reduce water wasting in IID's canals, in return for the saved water. It was the first glimmer of the sort of "water market" that economists had long dreamed of to improve the efficiency of California's water allocations, assembled through a very odd packaging mechanism that didn't look much like a market at all.

Rather than being on its heels fighting a losing battle, the Imperial Irrigation District saw the opportunity to become what one of the agency's leaders called the "new water market brokers of the Southwest."[18]

The initial phase of the deal looked straightforward: Los Angeles and its neighbors on the Southern California coastal plain would pay to line irrigation canals, a $260 million estimated cost at the time, in return for the saved water. But the community politics of Imperial County clouded the deal. Any effort to move water from the rural desert communities to the city raised reminders of the Owens Valley, where Los Angeles bought up water rights and moved water from a rural area in a way that still provokes bitterness a century later. In California, the words "Owens Valley" signify dangerous political ground. "Local people are understandably paranoid," one Imperial Valley water manager explained.[19]

For a time, paranoia won out over the opportunities to be the new water brokers, and Imperial Irrigation District leaders chose to fight rather than deal. The California State Water Resources Control Board in 1984 ruled in Elmore's favor, but the District launched a series of legal challenges that dragged the issues into the courts for most of the 1980s. In the end, the irrigation district lost, but by that time negotiations were well under way between the big farm district and Southern California water agencies that opened up room for a collaborative agreement that would benefit everyone. By 1989, the first agreement had been signed. It accomplished simple things like getting canals lined and automating irrigation gates, but also things that were more sophisticated, like payments by the Metropolitan Water District for efficiency improvements in IID's water-scheduling system. By 1990, water had begun moving from the Imperial Valley toward Southern California.[20]

The Sea Suffers

Despite the hard political slog, a series of agreements followed that further reduced water use in the Imperial Valley, moving ever more water to Southern California, especially to the San Diego County Water Authority. But eventually they bumped up against a very different problem from the one that brought John Elmore into court back in the 1980s. Instead of too much water going into the Salton Sea, by the late 1990s elevations began to decline, bringing the new problem of too little water going to the sea.

Any deal that brings efficiency to agriculture runs a hidden risk. What we think of as "inefficiency"—flood irrigation seeping down past a crop's root zone, or tailwater flowing out the bottom end of a field—is not water lost forever. Frequently it is water that plays a beneficial role wherever it goes next.

Most often, that role is aquifer recharge, with groundwater eventually flowing, hidden, back into rivers. Sometimes, tailwater ends up diverted into rivers directly, agricultural "waste" becoming a freshwater supply for downstream users. And in the Imperial Valley, that "waste" was what was propping up the Salton Sea. Drainage from Imperial Irrigation District farms has long flowed into the sea, in enormous quantities—an amount by some estimates equivalent to nearly 10 percent of the entire flow of the Colorado River. Reducing that "waste" of water has always seemed like just about the most important low-hanging water-conservation fruit in the Colorado River Basin. But water planners knew that if things in the Imperial Irrigation District got too efficient, the Salton Sea could simply dry up.

When Tina Shields began working for the irrigation district as a young engineer in 1992, the district was building dikes to hold back the rising water. By 2015, when she had risen to become the agency's water manager, one of her primary jobs was fighting the sea's decline.[21]

The "death" of the Salton Sea is a complicated problem. Some argue for simply letting it go, pointing to the accident of its creation in 1905 and saying it simply never should have existed to begin with. Writing it off completely could save a lot of water. But it provides one of the last remaining stops for migratory birds on that stretch of the Pacific flyway, invoking both environmental laws and values in any discussion over its future. More important is the health of the communities that surround it. Given the potentially toxic dust that would be left behind on barren salt flats, ready to be lofted by the next windstorm into a region that already has notoriously poor air quality, that was an unacceptable side effect for the people of the Imperial Valley.[22]

Thus the great dilemma posed by efforts to encourage water conservation in the Imperial Irrigation District to help reduce California's overuse of Colorado River water. Water conservation projects in the district, by reducing the amount of infiltration and runoff from the valley's farms, inevitably reduce flows to the Salton Sea.

When state officials agreed in 2003 on the "Quantification Settlement Agreement" needed to bring California into compliance with its 4.4 million acre-foot Colorado River allocation, it was understood that fallowing land and improving irrigation efficiency would reduce flows into the Salton Sea. If nothing was done, the sea's decline was inevitable. Fears about the impact of reduced flows to the Salton Sea threatened to blow up the entire deal. So to save the agreement, California state government made what to the communities of the Imperial Valley felt like a promise: you folks take care of the water conservation and transfer programs needed to reduce California's Colorado River use and transfer water from farms to cities, and we (state government) will take care of the sea. The promise included vague notions of an engineering fix, perhaps dikes to fence off a portion of the dry lakebed and a network of canals to preserve habitat and provide water to mitigate dust. It included some up-front money and a promise of studies to come up with a long-term fix.[23]

Imperial Irrigation District farmers did their part, reducing water use from 3.15 million acre-feet in 2002, the year before the "Quantification Settlement Agreement," or "QSA", as the deal is known, was signed, to 2.48 million acre-feet in 2015, a 21 percent reduction. It is by far the largest conservation project in Colorado River Basin history. But the state of California has failed to keep up its end of the deal, barely starting the steps needed to keep its promise of an engineering solution to make up for the loss of farm runoff to the sea. It is an example of the risks in large, multiparty water-management agreements. They depend on everyone keeping their commitments, and the state's failure created the risk by the early 2010s that the biggest water-conservation agreement in basin history could fall apart.

"We're pretty good at hiring attorneys over the QSA," Shields, the Imperial Irrigation District's water manager, told the audience at a water law conference in the spring of 2015.[24]

Productivity Keeps Rising

Imperial County farm-community members will tell you that all this cooperative conservation effort was undertaken with a gun to their heads, and they are right. Hovering over these agreements was always the fear that if the Imperial Irrigation District didn't deal, the federal government might try to come in and take the water by legal force, without benefit of compensation to the community. But in the face of that, the valley's agricultural community has shown remarkable resilience. Even as the water diverted from the Colorado River to Imperial County's farms declined, the region's farm income just kept rising. In 2013, the federal government estimated the county's total agricultural revenue at $2.3 billion, in inflation-adjusted terms 40 percent above what it was in the early 2000s, when water supplies were at their peak.[25] Water use was down 20 percent, yet farm revenues were booming.

How to explain this? It is an example of the point that agricultural economist Bob Young, talking about Arizona a half century before, made about the resourcefulness of farmers adapting to scarcity as their water supplies declined. They shift their cropping patterns, make more efficient use of water, and generally adapt such that farm income, even in the case of declining water supplies, keeps going up. In Imperial County, all you have to do is look at the lucrative vegetable business, the same thing that turned Yuma into such a powerhouse. From 1997 to 2012, even as water supplies were declining, total vegetable acreage went up by nearly 50 percent.

Acreage devoted to alfalfa, one of the lowest dollar-per-gallon crops, meanwhile, declined, dropping 25 percent in the five years after implementation of the Quantification Settlement Agreement began. Other low-value feed crops, such as bermuda grass and sudan grass, also declined.

Farmers know where to put their water to get the biggest economic bang per gallon.

CHAPTER 10

Emptying Lake Mead

IT WAS EARLY 2000 WHEN Terry Fulp saw the first glimmer of the problems to come. The hydrologist was part of a team doing the math on a proposal to change the way the federal government operated Lake Mead and Lake Powell, the two big reservoirs on western North America's iconic Colorado River.

In 2000, Lake Mead was full, water lapping at Hoover Dam's spillway gates. The full reservoir was a reassuring sight for the residents of the farms and cities dependent on the Colorado's supply. But gathered in a nondescript Southern California office park going over calculations with a team of technical experts, Fulp realized that things would not always be this way.

The team had been working "all hours of the day and night" on the final numbers needed for a federal report. As they sat down over pizza and beer one evening, one of the bosses asked a question: "If you could put something on a bumper sticker about what we've learned, what would it be?"

Fulp's answer was simple: "Lake Mead will go down." At a time when

Lake Mead, depleted by years of downstream water use (© John Fleck).

Lake Mead had been going up year after year, it was counterintuitive. But Fulp could see in the numbers a reality that would come to dominate his life, and the lives of Colorado Basin water managers and users, in the decade to come.

Fulp's group had been working with complex computer models used to simulate the operation of the river and its reservoirs as part of a major federal study, but Fulp did the calculation that finally convinced him on a sheet of paper.

Hydrologists call it a "mass balance calculation" and it's pretty simple, like balancing your checkbook: How much water do you already have in the system? How much flows in each year? How much flows out? Fulp started by scribbling on the piece of paper. "If we get 8.23 . . ."

Bound up in history, the number "8.23" is freighted with meaning. Under the river's operating rules, it is the required minimum delivery of water, in millions of acre-feet, from the basin's great upstream savings bank, Lake Powell, into the other, Lake Mead.

If you live in the states of the upper part of the basin—Wyoming, Utah, Colorado, and New Mexico—Lake Powell is your savings account. As long as those Upper Basin states have enough water in Lake Powell to release 8.23 million acre-feet of water each year, sending it downstream toward Lake Mead, they have met their obligation under the legally contentious web of rules known as the Law of the River. There is no consensus that "8.23" is a legal mandate, but as long that much water is moved downstream from the Upper Basin's savings account to the Lower Basin's, as a practical matter no one will put up a fight.

What Fulp's simple mass balance calculation showed was that 8.23 million acre-feet was not enough. The math is strikingly simple. California's legal entitlement is 4.4 million acre-feet. Arizona's is 2.8 million acre-feet. Nevada's is 300,000 acre-feet. Under international treaty, we are obligated to send 1.5 million acre-feet to Mexico. Add those up and you get 9 million acre-feet. Nine million is bigger than 8.23 million. Under normal operating conditions, the basin's books don't balance.

Fulp's calculation was a bit more complicated. There is extra inflow to consider from a handful of desert rivers that help add to the supply available in Lake Mead. But there is also evaporation and "system losses"—the basic inefficiencies in moving water through the complex plumbing of the Colorado River's Lower Basin.

Once Fulp added in those pluses and minuses, the problem looked even worse. "I was just a hydrologist in the middle of trying to figure out what these models were saying, and did we believe it, were the probabilities making sense?" Fulp told me in an interview fifteen years later.[1]

In the years since, Fulp has risen from "just a hydrologist" to director of the US Bureau of Reclamation's Lower Colorado River Region. If "Lake Mead will drop" has become one of the West's central problems, it is now Fulp's job to help solve it.

When Fulp told me this story, we were sitting in his office in the Bureau's regional headquarters, a grandiose white building atop a hill in

Boulder City, Nevada, surrounded by an expanse of lawn that is embarrassing in a desert city that averages less than six inches of rain a year. As he walked me to my car after talking for a couple of hours on a late winter day in 2015, we could look out over the great emptiness that has become Lake Mead. His bumper sticker slogan of fifteen years earlier had come to pass. Less than half full, Lake Mead's surface elevation had dropped a staggering 125 feet since 2000, enough water lost to have supplied nearby Las Vegas for the next half century. The dropping reservoir left behind a thick white scar they call the "bathtub ring," mineral deposits stretching to the high-water mark that offer a stark reminder of the problem.

The steps taken by water users downstream—cutting California's allocation, reducing Imperial's irrigated acreage, throttling back Arizona's groundwater pumping, steps that had seemed so hard when they were taken and that had proven so successful—had not fixed the problem. Water users downstream kept taking more water out of Lake Mead each year than flowed in, and Lake Mead kept dropping.

The Geography

As Lake Mead dropped in the years that followed, the arguments among Colorado Basin water users over interpretations of the river's operating rules had the look and feel of medieval scholars debating how many angels could dance on the head of a pin. But the stakes were enormous, because as they dithered, water withdrawals continued unabated and Mead's bathtub ring kept growing. If the problems were left unresolved, it was easy to see in Fulp's calculation that communities dependent on Lake Mead would quite literally run out of water.

The dispute hinged on complex, unresolved legal questions about the Law of the River. But more importantly, it exposed a moral and political rift: should Upper Basin water users, suffering under drought, be asked to pay more while California was allowed to suck surplus water from the system?

This is one of those times where the sometimes confusing geography of the Colorado River Basin matters a great deal. Water-management conflicts often revolve around upstream-downstream disputes, and the challenges in the early twenty-first century were an upstream-downstream geographical doozy.

There is an old saying in western water management: "I'd rather be upstream with a shovel than downstream with a decree."[2] The idea is that upstream water users have a natural physical advantage because they have the physical opportunity to take their shovel, dig a ditch, and put the water to use before it ever has a chance to make its way downstream. The great watershed that feeds the Colorado River's water into the river's deeply carved Utah canyon country has always appeared to offer that advantage, and for most of the twentieth century the Upper Basin states used it, digging ditches to move water to farms along the Rocky Mountains' west slope in Colorado, in valleys like Grand Valley, where the city of Grand Junction sits today.

To overcome the shovel/decree problem, the framers of the Colorado River Compact came up with a powerful rule. However much water those Upper Basin shovelers diverted into their ditches, they always had to leave enough in the Colorado River to meet downstream needs. The compact's framers set a number on that amount: 7.5 million acre-feet a year measured at a point just downstream from Lee's Ferry in a gorgeous red-rock canyon northeast of Flagstaff, Arizona. The water would then flow through the depth of the Grand Canyon before it was caught downstream behind Hoover Dam in the great reservoir known as Lake Mead.

The rule was written to allow some wiggle room—it really compelled the delivery of 75 million acre-feet over any rolling ten-year period, to allow for wet years and dry years in the highly variable Colorado River Basin climate. Over-delivery in a wet year would allow under-delivery in a dry year.

Construction of Glen Canyon Dam in the 1950s and '60s, just upstream from Lee's Ferry near the point where the Colorado River crossed from Utah into Arizona and entered the Grand Canyon, simplified the problem. Now surplus from wet years could be stashed in the giant reservoir known as Lake Powell, a water-banking savings account to ensure that the Upper Basin states would always have enough water to meet their 7.5 million acre-foot payment.

Thus each basin had its own great savings bank—Lake Powell for Wyoming, Utah, Colorado, and New Mexico, and Lake Mead for Nevada, Arizona, and California. Lake Powell's role was to store water needed to pay the Upper Basin's 7.5 million acre-feet per year bill to the Lower Basin. Lake Mead was a savings bank to store water needed for water users in Nevada, Arizona, California, and Mexico. Manage the savings accounts right, and there would be enough water to meet everyone's need when it was time to dig their metaphorical shovels into the ground and turn water into their ditches. But the rules were never entirely clear, and by the early 2000s, Upper Basin water users feared a downstream decree, in the form of legal action to enforce the 7.5 million acre-foot obligation, would trump their upstream shovels and ditches—so, as Lake Powell dropped, they would have to curtail their use to keep the full amount flowing past Lee's Ferry.

The Mexican Obligation

January 2000, as Fulp and his colleagues were finishing up their analysis of Colorado River Basin operations, would be the last time for a while that Lake Mead lapped at Hoover Dam's spillways.[3] The decline of the great reservoir began in the summer of that year. By 2015, Lake Mead had reached the lowest level it had been since the Bureau finished Hoover Dam and began filling it in the 1930s.

The problem, as Fulp saw so clearly, is simple: downstream water use is greater than the supply that nature and the upstream water users provide. But after a big policy push in 2002 and '03 to try to curtail California's overuse of Colorado River water, the years that followed settled back into a predictable debate: not over how to use less water, but rather over the rules that decide who takes the hit when shortage sets in.

While the California experiment, cutting that state back to its 4.4 million acre-feet per year allotment, seemed successful, the next steps proved substantially harder. No one else wanted to take on the burden of using less water, which in the end is the only way to keep from draining the basin's reservoir supplies. Everyone was trying to find a way of reading the rules so that, as scarcity set in, they wouldn't have to take a water-supply hit.

To grasp the legal and political debates among basin water users in the years that followed requires a deeper understanding of the rules that lay behind that seemingly sacrosanct Upper Basin 8.23 million acre-foot delivery requirement at the heart of Fulp's calculations. That number depended on a particular interpretation of the river's operating rules, and as Lake Mead and Lake Powell continued to drop, that interpretation was called into question.

You need to invoke two separate rules to get to a total of 8.23 million acre-feet. The first is the Upper Basin's legal requirement under the Colorado River Compact to deliver a minimum average of 7.5 million acre-feet per year of water past Lee's Ferry. But to get up to 8.23 million acre-feet, you also needed to assume that the Upper Basin is required to deliver extra water above and beyond that minimum obligation to provide a share of the water delivered to Mexico under a 1944 treaty between the new nations. That number was buried in four pages of turgid prose that only a water lawyer could love: the 1970 federal regulation known as the LROC—"Criteria for Long-Range Operation of Colorado River Reservoirs."

In 2004, the states of the Upper Basin did not like the way the LROC (they pronounce it "el-rock") was being used. They balked at the idea that they were required to deliver a share of the Mexico water. In the current state of the river's management, the representatives of the four Upper Basin states wrote, "the Upper Basin has no obligation in this regard."[4]

If they were right, it would make the math problem identified by Fulp four years earlier even worse. With 7.5 million acre-feet delivered instead of 8.23 million acre-feet, Lake Mead would go down that much faster. To the states of the Lower Basin, it was as if a business partner with a $1.5 million annual partnership declared that he had no obligation to pay his share of the debt, a trio of Arizona water lawyers wrote.[5]

Any attempt to decide what's really "fair" in distributing the basin's water is a doomed exercise, but there is one crude, easy-to-see metric: how much water is in Lake Powell? How much in Lake Mead? The two reservoirs, book-ending the Grand Canyon, act like savings accounts for the two basins, and if one is empty while the other is full, that just looks unfair. By 2004, the balances clearly looked unfair.

Bad drought years—2002 was the worst on record—had left the upstream reservoir, Lake Powell, so enfeebled that the National Park Service abandoned the marina at Hite, in the red-rock canyon country at Powell's upstream end. What once had been lake at the site of the marina was turning rapidly into narrow river channel as Lake Powell dropped away. Workers towed the floating gas station and marina into deeper water, leaving behind a forlorn concrete ramp stretching into what used to be a recreational boaters' paradise but by the early 2000s was little more than a muddy former lake bottom.[6]

Lake Mead was dropping, too—its surface elevation down 70 feet in the first five years of the twenty-first century. Together, the combined water in storage in the two reservoirs had been cut nearly in half.

Meanwhile downstream, water users watched demand skyrocket. The population of Las Vegas grew 23 percent from 2000 to 2005, and Phoenix grew 18 percent. Water managers were nervous, but as they watched the two reservoirs drop in tandem, they had the feeling that they were all in this together. There were feuds, but the shortage was primarily treated as a shared problem.

One of the first obvious signs that it might not be such a shared problem came in 2002, the year Hite was abandoned. The lowest snowpack in history left the Upper Basin parched. Farmers in Utah and Wyoming estimated that they suffered a 337,000 acre-foot shortfall.[7] With less water in the snowpack in the mountains above them, they had no choice but to use less water. Meanwhile downstream, there was so much water in Lake Mead that the US Bureau of Reclamation declared a surplus. California got 875,000 extra acre-feet of extra water in 2002, 20 percent above the state's normal-year legal entitlement. Lake Mead was dropping, too, but Lake Powell was dropping more.[8]

To Upper Basin water users, this seemed unfair, and they thought they understood the cause. The states of the Upper Basin argued that their reservoir was being lowered so quickly because they were being overcharged—that they were being made to contribute too much water to meet US delivery obligations to Mexico.

Under existing interpretations of the Colorado River Compact, it had been generally agreed that each basin was responsible for half of the United States' treaty obligation to deliver 1.5 million acre-feet per year past the US-Mexico border at Morelos Dam. But water managers in the Upper Basin had always chafed at the agreement, harboring their own legal interpretation that, when there was "surplus" in the system, the entire 1.5 million acre-feet should come from that surplus. Instead, surplus water had for years been sent to California while the Upper Basin's dwindling bank account in Lake Powell was being charged to meet the US obligation to Mexico. Instead of declaring a "surplus" and sending the extra water to California, the Upper Basin argued, the extra water should have been sent to Mexico, allowing more water to be held back in Lake Powell as a savings account for the future.

This is one of those problems that did not come up for most of the history of the management of the Colorado River because there had long been enough extra water for everyone. But much in the same way that the basin states were forced to grapple with California's excess use as demand finally rose to meet supply, or deal with the loss of water because of the salty water flowing from Wellton-Mohawk in Arizona, they were now forced to grapple with the issue of who bore what responsibility for the Mexican allocation.

"They've never had to face a shortage of this consequence," Las Vegas water chief Pat Mulroy said at the time. "When you're right up against it and facing the possibility of inadequate supplies to municipalities or farmers or jeopardizing recreation values, these are very tough choices."[9]

With Lake Powell dropping, the Upper Basin saw a need to protect "their" water. The ability of Glen Canyon Dam to generate power for some 200 mostly rural communities was at risk. If the drought continued to sap the basin's snowpack at the rate it had been, Glen Canyon Dam would be unable to generate electricity at all by February 2006.[10] Upper Basin agriculture also was being hit hard by the drought.[11] With demand across the basin growing, "it is apparent these issues will not go away, even if we are blessed with a few years of favorable runoff," the Upper Basin representatives wrote.[12] The states were at an impasse, and there was a very real chance that the whole thing would end up in court.

Most of the water managers involved were too young to have lived through the cloud of uncertainty that hovered over the basin the last time such a dispute had ended up in court, when the US Supreme Court worked from 1954 to 1963 to sort out the *Arizona v. California* litigation. But they had internalized the lesson from that era, that courts are a lousy way to deal with questions like this. "That route takes the decision out of the water managers' hands," explained John Entsminger, one of Nevada's representatives in the negotiations that followed. "Are we going to let guys and gals in black robes start making these decisions for us? Or are we going to come together and maybe not have a perfect solution from everybody's perspective, but a solution that works for everybody?"[13]

The Network

The group of people who gathered to work on this issue lacks a formal name. John Entsminger, the Colorado-born lawyer who rose to head Las Vegas's largest municipal water agency, is one of the people who calls it simply "the network." The existence and functioning of the network illustrates the reality of problem solving in a river basin where water crosses borders, where it must be shared, but where no one is in charge.

Though there are some formal structures through which it operates, the network is not a formal thing. It includes representatives of the seven US Colorado River Basin states and the major water-using agencies within those states. A team of federal lawyers, hydrologists, and water managers are central players. It also includes a handful of outsiders—attorneys and representatives of environmental groups who have learned the lingo and earned the trust to participate in the discussions.

They are people, explained Anne Castle, a water attorney, former assistant secretary of the interior, and one of the network's key participants, who are "laser-focused" on the reality that in order to meet their own states' and water agencies' needs, the basin's broader problems need collective action toward collective solutions.[14]

You have to have formal institutions—federal and state agencies, formally established working groups reaching across boundaries, and officially designated negotiating teams, all operating under written rules. But for those institutions to function well, you also need informal relationships, across institutional boundaries, among people who represent different communities and interests, yet understand one another's needs and share common values.

These people meet regularly. It has not been uncommon, when working on this book, for me to bump into more than one of them in the halls of a hotel in Las Vegas, Nevada, or Santa Fe, New Mexico. When the network gathers in Yuma, Arizona, I learned, it's a good bet that you can find the people you want to talk to in the lobby or out by the pool at the Hilton Garden Inn on the banks of the Colorado River.

Researchers who study these formal and informal institutions talk about "social capital"—"the shared knowledge, understandings, norms, rules, and expectations about patterns of interactions that groups of individuals bring to a recurrent activity," in the words of Elinor Ostrom.[15] The word "capital" is chosen carefully, suggesting that its role is every bit as important as the physical capital: the dams, pumps, and canals through which the water moves.

It was "the network" that ultimately put together the 2002 deal to reduce California's water use, and by 2005 it was being put to a test. Somehow, the network's members now had to come up with a new set of rules that could both balance reservoir levels in Mead and Powell, as well as provide some certainty for how shortages would be handled among the Lower Basin states if Lake Mead kept dropping.

It meant understanding one another's positions, but it also meant honoring a shared goal—keeping their dispute out of court. "We knew where the disagreements were," said Entsminger, "and the choice was litigate those disagreements or find a working solution."[16]

The Negotiations

In the two years that followed, there were times when Terry Fulp thought the whole thing would fall apart. Fulp had risen to be deputy director of the Bureau of Reclamation's Boulder City office by that point, and he was given the unenviable task of trying to shepherd the negotiations to come up with new operating rules for the basin. "I had many nights that I went home late at night and sat there and thought, 'We're not going to make it,'" he said.[17]

Just as in the close-but-not-quite-there moments in 2001 and '02 over the California allotment, a deal seemed within reach. The biggest hurdle was the need for an agreement on reduced allocations to downstream water users as Lake Mead dropped. California wanted deep cuts in the amount of water released each year to Lower Basin water users, knowing that its priority status would mean the burden of those cuts would fall entirely on Arizona and Nevada. Arizona pushed back, and the states agreed to more-modest cuts that were not deep enough to completely halt the drop in Mead, but were a start.[18]

But the details—how much to release from Lake Powell each year, and under what circumstances—increasingly seemed insurmountable. Rules that left enough water in Lake Powell to make Upper Basin states comfortable left Lower Basin water users, especially in Arizona, nervous that Lake Mead would drop to dangerously low levels. By the summer of 2007, the network was staring failure in the face. They had gathered in Phoenix (they picked the site to coincide with the wedding of Colorado negotiator Scott Balcomb's daughter) when they awoke to a newspaper headline announcing that Arizona wanted to pull out of the negotiations. "We worked hard to try to put together this agreement, and the only thing Arizona asked is that it did not harm Arizona water users," said Herb Guenther, director of the state department of water resources, to the *Arizona Republic*.[19]

Said Entsminger, "Everybody was blindsided."[20]

With the possibility for agreement in tatters, a meeting the following month at Bishop's Lodge outside Santa Fe offers one of the clearest examples of how the network operates.

Bishop's Lodge is a historic site in Colorado River Basin history, the place where the original version of the network had gathered seventy-five years earlier to finalize the Colorado River Compact. In recent decades, a nonprofit called the Water Education Foundation has organized an invitation-only return to Bishop's Lodge every two years to bring together senior basin water managers. It has become one of the most important gatherings of the network. There are seminars and talks, but more important are the full breakfasts before the day begins, the hallway conversations during the day, and the happy hours after. It is the place where network governance happens.

It was there in 2007, in a bar after a day of meetings, that Nevada hydrologist David Donnelly hatched the scheme crucial to saving the agreement. Donnelly's plan offered a way, under certain circumstances, to release a bit of extra water from Lake Powell to bolster Lake Mead.

Donnelly's breakthrough required two things that are critical to the success or failure of the process. The first was Donnelly's deep understanding of how the system itself works, the arithmetic of water stored in one dam moving downstream to be captured by the next. But more importantly, the idea evinced a deep understanding by Donnelly, a Nevada representative, of the other states' needs. If they failed to come up with a deal, Nevada arguably had the most to lose. If they couldn't come up with rules to slow the decline of Lake Mead, the Las Vegas water-intake pipes were vulnerable, posing the risk of a city going without water. So for Nevada, the most important thing was a deal that could get the other six states to sign on. "Nevada had a lot of skin in the game," Donnelly told me. "Our main goal was to try to get everyone else on board."[21]

They succeeded.

The Deal

When it was finally signed in December, the resulting agreement was historic. For the first time since the Colorado River Compact was signed in 1922, the states on the US side of the border had agreed that there might not be enough water to cash the check written in the compact—that there might be times when users would simply have to do with less.

Under the deal Arizona and Nevada agreed that when Lake Mead dropped to a surface elevation of 1,075 feet above sea level, each would take a cut in their guaranteed annual delivery. When Mead hit 1,050, the cut would be bigger. At 1,025, bigger yet.

Never before had Colorado River water users formally agreed that shortage was a reality—that full deliveries year in and year out were not an immutable right. The agreement also provided new flexibility for managing the system, allowing major water users to conserve water without losing it, banking the savings in Lake Mead for use in later years. No longer was "use it or lose it" the official operating policy of the Colorado River.

But you can see the deal's poorly stitched seams—choices that were necessary to keep it together, but that weakened its ability to truly address the problems it was meant to address.

Under the deal, when Lake Mead drops to a surface elevation of 1,075 feet above sea level, Arizona and Nevada agreed to take less water from the reservoir. The purpose was to slow the decline, but the shortages are modest—an 11 percent cut to Arizona's allotment, and a 4 percent cut to Nevada's. California, the victor in the battles for priority rights in the 1960s, still gets its full allocation.

When the negotiations began, some of the participants, led by California, argued that far deeper cuts in usage would be needed to preserve the water supply over the long run. But California's proposal came with a poison pill that made bigger cuts a nonstarter: the reductions would fall entirely on Arizona and Nevada. California thought that water conservation was critical, but that it was its neighbors' responsibility to do it.

By the summer of 2014 it was clear that the deal, historic as it was, had not gone far enough. By Arizona's calculations, even with the new rules, and absent a string of wet years, Lake Mead could be unusably empty by the early 2020s. They had slowed the inexorable arithmetic of Fulp's prophetic calculation, but they had not stopped it. The network had more work to do.[22]

CHAPTER 11

Who's Left Out?

THE LANGUAGE OF THE 1922 Colorado River Compact regarding Indian water rights feels like an afterthought: "Nothing in this compact shall be construed as affecting the obligations of the United States of America to Indian tribes." Native Americans went unrepresented in the compact's negotiations, and their water rights were left hanging. They were treated not as communities whose needs mattered, but more as a problem to be dealt with later.

Nine decades later, as the basin's water managers came together to embark on their sweeping Basin Study, the most ambitious effort yet to grapple with the Colorado's problems of supply and demand, it seemed that little had changed. As the steering committee was formed to shape the direction and substance of the study, Native communities were still unrepresented in a process that, in the words of the Jicarilla Apache Nation's Darryl Vigil, "could significantly and adversely impact tribal water rights and tribal usage of water."[1]

Relationships have been the common theme in our water success stories. Whether dubbed "the network," or "institutional plumbing," or simply human connection, formal and informal relationships among institutions and often among specific people lie at the heart of problem solving. We have seen, for example, how Bennett Raley's decision to include environmentalist Jennifer Pitt on a 2004 Grand Canyon river trip eventually led to a breakthrough in dealing with environmental problems in the Colorado River Delta, and how brainstorming by Nevada's David Donnelly opened breathing room to satisfy Arizona's concerns about the 2007 shortage-sharing agreement.

You can think of these relationships as the dividends of an investment in "social capital"—an investment that is every bit as substantive and important as those made in physical facilities—the dams and canals, the pumps and pipes—that move the water. Much of it is an investment in the tools of civil society, the institutions (both formal and informal) that emerge to handle the tasks. Like physical capital, it takes time and resources, and it decays with age.

But social capital can also create barriers to success, raising questions of equity and justice as we adapt to scarcity: namely, who is left out of these processes? Who was *not* invited on Bennett Raley's river trip, whose interests were therefore not represented? Which important stakeholders lack the resources to participate in the time-consuming and expensive processes through which decisions are made? How do we avoid turning this into management by and for a powerful elite, to the exclusion of those who aren't members of the in group?

At times, the problem amounts to overt discrimination. At others, there is simply confusion about whom these decisions will affect. This includes electricity consumers who might be impacted by a change in operations at one of the basin's big power-generating dams, as happens at Glen Canyon Dam, the Upper Basin's big storage reservoir, when more water is sent through the Grand Canyon to enhance environmental flows. It includes the public health community in the Imperial Valley, where air-quality problems could result from changes to water management around the Salton Sea. The issue here is not so much barring people from participating in the process as it is finding the right people to invite in. How do we decide where to draw the boundaries around the problem we are trying to solve in order to ensure that all the important stakeholders are included in the decision?

The history of the last two decades of problem solving on the Colorado River suggests progress is being made on this issue, especially in including environmental interests and the nation of Mexico in the decision-making process. But over and over, those trying to sort out the Colorado River's problems find that they've drawn the boundaries at the wrong place, and that something done within the river-management community has an impact on some group, interest, or issue that has been left out.

Whitehorse Lake

On a blindingly sunny January day a few years ago, I stood outside Chee Smith Jr.'s house as he opened the spigot on a new faucet poking up in his yard. Quickly, the president of the Navajo Nation's Whitehorse Lake Chapter shut the valve. The high desert community is far from pretty much everything. It is, for example, an hour's drive from Grants, New Mexico, the city with the nearest laundromat. That matters because Whitehorse Lake for years has also been far from the nearest reliable supply of water. The new flow, from the first running water his little community has ever had, was precious. "Water is gold out here," Smith said. "It's our life."[2]

People argue about what might be considered the most water-stressed community in the United States, but Whitehorse Lake and other Navajo chapters on the fringes of the Indian nation's sprawling 27,000-square-mile reservation are certainly contenders. In all directions—Phoenix to the southwest, Albuquerque to the east, Farmington to the north—non-Indian communities were built and prospered on the back of Colo-

Water barrels behind a sheep pen at the Navajo community of Whitehorse Lake (© John Fleck).

rado River Basin water, subsidized by the nation's taxpayers. But White-horse Lake, and much of the Indian world that surrounds it, was left out.

Per capita water use on the Navajo Nation is half of that in neighboring non-Indian communities, in large part because 40 percent of Navajo homes lack full indoor plumbing, defined by the US Census Bureau as running water, a shower or bathtub, and a toilet. The comparable number for the United States as a whole is 2 percent. You can't use water that you don't have. Instead, those Navajos without plumbing haul water, filling big tanks on the back of their pickup trucks, a process that comes at a staggering cost in time, money, and health (because of water-quality issues) that their non-Indian neighbors don't have to face.[3]

Whitehorse Lake got its name from water that used to pool behind a hand-built earthen dam, long since gone. Similarly, there are no lakes at the community of Seven Lakes, down the road. But the names show

how much water matters in this largely treeless high country stretching north from the sacred Mount Taylor—Tsoodził, the turquoise mountain. New Mexico State Route 509, the two-lane paved road from the south that is the main connection to the outside world, crosses the barely noticeable crest of the Continental Divide just south of the Whitehorse Lake Chapter House. As a result, the community falls squarely within the Colorado River Basin—a crucial fact for its future.

Averaging less than nine inches a year of precipitation,[4] this is far from the driest place in the United States. Though it technically counts as a desert, the Colorado River and its tributaries pass through places that are far drier as the river system drops from the high country of the Rocky Mountains westward through the great desert lowlands on its way to the sea. But the residents of Whitehorse Lake and many other communities scattered across the Navajo nation have three strikes against them that other Colorado Basin communities facing similar precipitation shortfalls do not have: they lack an accessible supply of groundwater; they are far from the Colorado River or any of its tributaries; and they lack the affluence and political power needed to overcome their communities' geographical and hydrological shortcomings.

Most houses scattered across the Whitehorse Lake community have an outhouse and, less noticeably, a collection of buckets and barrels for hauling water. Residents drive to the nearby chapter house to fill the buckets at the community's small potable water well for indoor water use. Windmills that pull lower-quality water from the ground are used to fill larger barrels that are carted home for the livestock, often goats, that are central to the Navajo community way of life. But it's no secret that people sometimes drank that water too, lower quality or not.

By virtue of having been here first, the Navajo are, on paper, among the most water-rich people in the region. But if, as Smith said, water is gold out here, the Navajo Nation has had a hard time cashing in. For decades, the water-management community has paid lip service to honoring an obligation to the Navajo, but in practice the big, expensive, federally subsidized plumbing has channeled the water elsewhere.[5]

In November 2012, twenty-four homes in the Whitehorse Lake community got their first running water. In December 2013, water service was extended to another twenty-one, including the home in which Chee Smith Jr. grew up. Down the road from Smith's house is a rickety mobile home with a basketball hoop where his niece now lives. She moved back to the ancestral homeland from an off-reservation city, drawn by the fact that the community finally has running water.[6]

Indians and the Basin Study

The uneasy relationships between Native communities and the federal bureaucracy are painfully apparent in the Bureau of Reclamation's "Colorado River Basin Water Supply and Demand Study." The study, completed in 2012, represented the first comprehensive effort to reconcile the Colorado River's dwindling supply in an age of climate change with the increasing demands of basin water users. As such, it marked a major milestone in coming to grips with scarcity. But from the beginning stages of the study's planning, Native communities—who hold huge but in many cases unquantified rights to Colorado River water—were left out.

It is the latest episode in a long and ugly history. In the allocation of the West's water, Indians have often been left behind—pushed off their land and onto reservations, often violently, and then left without enough water to use the land set aside for them.

On paper, their rights to water have long been well established. In 1908, the US Supreme Court ruled in the case of *Winters v. United States* that the federal government had an obligation to protect the water rights of the Native community of the Fort Belknap Indian Reservation, which were established when land for the reservation was set aside in the winter of 1887–88. Anglo settlers diverting water upstream could not deprive the Indians of their due, the court ruled.[7]

The Winters decision included a critical precedent by sidestepping the state-based doctrine of prior appropriation. Unlike the settlers, who had to put water to beneficial use to claim priority rights, the Indians were presumed to have the rights to the water they needed to make use of their land at the time the reservation was created, even if the water had not yet been put to full use.

This has left significant uncertainty hanging over water-rights allocation throughout the Colorado River Basin ever since—an uncertainty largely ignored in the rush to develop the dams and canals needed to put the water to use.

For the first half of the twentieth century, the *Winters* decision "lay dormant," to borrow the odd language of the lawyers.[8] That's a fancy way of saying that, as the vast Colorado River plumbing system was being built across the West, Native American rights to water, and the plumbing needed to deliver it, were largely ignored.

The question of Native water rights lurked in the legal background until the US Supreme Court addressed it in the 1963 decision that it handed down in the case of *Arizona v. California*. While much of the public attention focused on the settlement of the conflict between Arizona and California, the court also advanced the rights of Indians to water in ways that continue to reverberate half a century later. The *Winters* decision established Native American rights to water, but it left unsettled the question of how much.

Simon Rifkind, a lawyer tasked by the court with sorting out the mess, suggested an answer that would extend the *Winters* decision. If the goal of establishing Indian reservations had been to create agricultural communities, Rifkind argued, then the Indians needed enough water to do it. Arizona had argued that the Indians' water rights should be narrowly construed, matching the amount of water to the number of people on the reservation. But Rifkind's conclusion was much broader: "The magnitude of the water rights created by the United States is measured by the amount of irrigable land set aside within a Reservation, not by the number of Indians inhabiting it."[9] Rifkind's recognition of the potential future needs of the community, rather than simply those of the people there at the time the reservation was formed, was crucial. "It must have been apparent," he wrote, "that unless the United States reserved water for the land at the time of withdrawal, there might be no water left to appropriate at the time that the land was needed for the purposes for which it was withdrawn."[10]

The Supreme Court bought Rifkind's argument, and then extended it in an important way. Whether the court meant this extension or not is a subject of debate, but the court's muddy language indicated that, while agriculture may have been the federal government's reason for establishing some reservations, it need not be the only one. Native Americans, the court's ruling seemed to suggest, might be entitled to enough water for other things as well.[11]

But while in principle the decision in *Arizona v. California* seemed to substantially advance Native American water rights, the practice turned out to be more difficult. The process that led to the *Arizona v. California* decision also demonstrated the extent of the problem. Repeatedly, the tribes attempted to formally intervene in the proceedings to defend Native water rights. The court repeatedly refused to give them standing.[12] In what is arguably the most important forum for Colorado River Basin decision making, Native communities were once again left on the outside, barred from formal entrance to the process.

Despite the tribes' absence, the court allocated important water rights to some tribes. The court, for example, awarded the Colorado River Indian Tribes senior rights to enough water to farm 108,000 acres of rich Colorado River bottomland, primarily on the Arizona side of the border south of the town of Parker. The land today is a blanket of green, dominated by alfalfa, cotton, and durum wheat, with a sprinkling of fruit and winter vegetables.[13]

The Colorado River Indian Tribes are among twenty-two tribes that have had their rights quantified either through court action in the *Arizona v. California* case or negotiated settlement in the years that followed. But there are twelve tribes whose water rights remain undefined and are likely not being met with current supplies. The result is an uncertainty that, in the words of Native legal scholars Amy and Daniel Cordalis, has "left an unwanted cloud over the Colorado River Basin." It leaves the tribes, among the poorest communities in the nation, without the water they need to strengthen their communities and economies, and leaves non-Native water users unsure about how much may eventually be diverted for Indian use.[14]

Among those still lacking clarity about their rights, and water for their people, is the Navajo Nation. In 2009, Congress ratified an agreement allocating a big chunk of Colorado River water to Navajo land in northwestern New Mexico, water that would come from New Mexico's share of the river's Upper Basin allocation. It was that agreement that also finally brought Chee Smith Jr.'s community of Whitehorse Lake and others like it in the arid, rural country under the same umbrella of federally funded water infrastructure that non-Indian communities across the Colorado River Basin have long enjoyed. But litigation quickly followed, as non-Indian water users complained about the deal.[15] And because water rights are allocated state by state, the Navajos' claims in Arizona still remain unsettled, leaving little water and great uncertainty for members of the Navajo Nation within Arizona, including those living on the reservation's northwestern fringe, communities every bit as water-short as Whitehorse Lake.

Mexico

The second major constituency left out of Colorado River networks for much of the last century has been the nation and people of Mexico. The negotiations that enabled the 2014 environmental pulse flow through the Colorado River Delta marked an important step toward solving this problem. But this cannot erase a long history of exclusion of one of the most important groups of stakeholders on the river. Nowhere was this problem more clearly on display than in the All-American Canal.

Stretching from Imperial Dam to the Imperial Valley in the southeastern deserts of California, the canal passes through great fields of sand dunes. Unlined through most of its history, the canal had always leaked. Adding a concrete lining had long been seen as a way to save water and reduce the allocation flowing to Imperial without reducing acreage being farmed. But lining would come with a price. The canal seepage was refilling an aquifer that flows south into Mexico, where Mexicali farmers had come to depend on the flow. The seepage also fed some of the few wild wetlands left in the delta region.

The decision to line the canal was far easier because an international border, and the history, politics, and law associated therewith, blocked one of the most important impacted parties from having a seat at the table when the deals were being made.

Canal lining has long seemed like the lowest-hanging fruit in Colorado River Basin water-conservation management. The network of massive unlined canals that move water through the desert from the river to the farmers of the Imperial and Coachella valleys always leaked. In the 1970s, this was the first place water managers looked to squeeze some extra out of the system to meet overall delivery obligations, lining a forty-nine-mile stretch of the Coachella Canal, which carries water along the northern edge of the Salton Sea to the date orchards, citrus groves, and golf courses of the Coachella Valley.

This made sense. Seepage from the canal was essentially lost for human use, mingling with high-salinity groundwater and eventually ending up in the Salton Sea. But the next major canal-lining project was not so clear in terms of winners and losers, and the losers—Mexican residents of the Mexicali Valley—were excluded from the decision-making process.

In 1988, Congress authorized lining the All-American Canal. The deal came as part of a larger package aimed at settling water rights and federal obligations in the region, but because it was US legislation aimed at US water management, it left out the Mexicans entirely. Formal discussions were launched between the US and Mexican governments, but they were largely limited to the US government's insistence that the seepage had always been US water, to which the users across the border in Mexicali had no legitimate claim.[16]

This did not sit well. While the US effort to find the funding dragged on, the conflict festered. The US government did additional reviews of the project's environmental impacts, but refused to consult with the Mexican government.[17] On the Mexican side of the border, public opposition rose all the way to the nation's president, Felipe Calderón, who said during a speech in Tijuana that the canal lining would "cause enormous damage to the environment and the economy of the Baja California border."[18] In 2005, as construction was finally set to begin, a coalition of US environmental groups and Mexicali Valley civic organizations sued.

They lost, but the US approach to the action "drove the US-Mexico border water relationship to a new low," wrote US water attorneys Jonathan King and Peter Culp and Mexican researcher Carlos de la Parra, "and provided an excellent example of the resentment that the continued, arms-length water-management relationship could generate."[19]

It was not the first time the Mexicans had been left out. As we saw earlier, for example, solving the problems of the Wellton-Mohawk Valley required a callous willingness on the part of US water managers to dump salty water onto the irrigation and municipal intakes of their southern neighbors. Similarly, attempts to conserve "wasted" water by building temporary reservoir storage near the US-Mexico border to capture water that would otherwise be spilled from the Colorado River itself into Mexico came at the expense of downstream natural habitat that eked out a meager survival on the surplus water.

The end of abundance was forcing US water managers to tighten up the operation of their system, so reducing the "waste" of water made sense. But one person's waste was another's bounty, and all the losers in this zero-sum game were on the southern side of the US-Mexico border, left out of the discussions.

Environmental Values and the Surplus Guidelines

I've described the negotiations that led in the early 2000s to reductions in California's Colorado River water use as a success, and—measured by the standards of those driving the process—they were. The old "water buffalos," the representatives of the agencies responsible for moving water out of the Colorado River and putting it to human use, dominated the discussions and defined the goals. By their criteria, creating a framework to address the problem of California's overuse of Colorado River water was astoundingly successful. But the only way to get the deal's hydrologic equations to balance was to sidestep environmental issues in a process that illustrates the shortcomings of insulated negotiations. If you can only solve your problem by cutting water to stakeholders who are excluded from the process, you've got a serious shortcoming, and the resulting side effects are likely to linger. That is what happened with environmental interests in the 1990s, and some of the problems created then are still around today.

You can see the failure in an exchange of correspondence in 2000 between a coalition of environmental groups and the Bureau of Reclamation. It began with a plea in February of that year from the groups that the Bureau's new river rules should not ignore water for the desiccated Colorado River Delta. For years, the groups had been pushing without success to improve natural habitat in the delta. Now, the Bureau of Reclamation was contemplating changes in its river management that had the potential to make the already bad delta environmental situation even worse.

To meet the delicate balancing act of providing a "soft landing" for California, the new Bureau plan would capture the last bits of "surplus" that had occasionally still flowed down the river channel to the delta, choking off what little water was left and diverting it as well. The environment was left bearing the cost of California's overuse.[20] The groups reminded Bureau officials of the tremendous amount of Colorado River Delta habitat that had already been lost as upstream rivers choked off the Colorado, diverting the vast majority of the water to farms and cities in the United States.

In light of what followed, the environmentalists' February 2000 letter looks like a hopeless, strident, and naive demand. "Environmental needs must be met before any quantity of discretionary water is dedicated to consumptive users. Until then it is not truly 'surplus,'" the groups wrote.[21]

The Pacific Institute's Michael Cohen, who has become a quintessential example of the environmentalists now admitted into the inner sanctum of Colorado River Basin policy debates, laughed when I reminded him of the combative tone of the letter that he and his colleagues had written more than a decade earlier. "We were young," he said.[22]

The "we" consisted of the most prominent environmental groups on the US side of the border, including American Rivers, the Environmental Defense Fund (then known as "Environmental Defense"), and the Sierra Club. They offered up a proposal that, they argued, could meet California's needs while at the same time freeing up water for environmental flows in the river's channel in the Colorado River in Mexico. Their proposal included a strict requirement: that the US secretary of interior would not be permitted to declare any waters surplus and available for human use in California or elsewhere until there were scientific assurances that there would be enough water to maintain environmental flows in the delta.

The federal government would have none of it, arguing that it had no responsibility for the environmental impacts of its actions in Mexico. "The delivery of surplus water to Mexico is beyond the purpose and need for interim surplus criteria," the Bureau of Reclamation wrote in a formal response to the proposal.[23] But by the time the official response had been issued in late 2000, it was already too late. A smaller group of US environmentalists, joined by colleagues across the border, had already filed suit, charging that US water managers were ignoring the environmental impacts on the Colorado River Delta of Mexico.[24] Excluded from the decision-making process, the groups saw no alternative but to take the matter to court. They lost. The basin's water managers seemed free to ignore this critical constituency.

There are other cases like this. The failure to include communities that could be harmed by the decline of the Salton Sea may be the most important. Conflict between water managers and the electric-power community over changes to the operation of Glen Canyon Dam is another.[25] All these cases raise questions of environmental justice and suggest that water managers need to find ways to make their efforts more inclusive. But even if you ignore those moral questions, exclusion poses a risk. Parties left out, who are harmed by decisions made by insular circles, can derail important efforts to solve the basin's problems.

CHAPTER 12

A Beaver Returns to the Delta

DANIEL TREMBLEY MACDOUGAL, boating through the Colorado River's great delta in 1905, found a landscape incongruous to a New Yorker—in the midst of a great desert, a jungle "sufficient to support a vast amount of native animal life." Spread over some 3,000 square miles, the delta seemed impervious to the forces that were reshaping the landscape of western North America. He wrote: "The countless millions of young willow and poplar shoots supply food for the beaver, which bids well to hold out long in the impassable bayous and swamps against the trapper foe."[1]

The landscape today would be unrecognizable to MacDougal. Scraggly salt cedars, a Eurasian invasive, have taken the willows' place, flanking a largely dry channel, more desert wash than great American river. But as one of the greatest challenges to environmental river management in the United States and Mexico, that dusty river channel also offers one of the greatest signs of hope.

Two hydrologic facts shaped the landscape that MacDougal described more than a century ago. The first was the tremendous volume of water the Colorado River offered up as it spread across the desert before reaching the Sea of Cortez. Aldo Leopold's poetic description of the river as it slowed in its last miles through the labyrinth of its distributary delta is often quoted. He and his brother had canoed the delta in 1922, before upstream dams changed it forever—"a hundred green lagoons" teeming with life: "For the last word in procrastination, go travel with a river reluctant to lose his freedom to the sea."[2]

The second hydrologic fact was the annual cycle, the rise and fall of the river as it flooded with the runoff from the Rocky Mountains' melting snow, then dwindled in the quiet of winter. MacDougal described water "a height of a hundred feet above low-water mark" by mid-summer as snowmelt submerged the landscape. Some 70 percent of the river's entire flow would pass through the delta between May and July. James Ohio Pattie, an early beaver trapper, described the river bottoms as being six to ten miles wide and "subject to inundation in the flush waters of June."[3]

The trappers had made token efforts. One of the first English-language accounts we have of the delta comes from Pattie, who tried his hand there in the 1820s. Floating down from the Gila River in current-day Arizona into the delta, Pattie and his party caught so many beavers that they had to build an extra canoe to carry their haul.[4]

When MacDougal saw the beavers, he imagined "the trapper foe" to be their biggest enemy, never considering the real threat to the beaver—that the river itself might disappear. Yet by the 1960s, with the completion of Glen Canyon Dam, flows to the delta dropped to nearly nothing. Only in times so wet that the great reservoirs upstream "filled and spilled"—for a few years in the 1980s, and again in the late 1990s—did water make it past Morelos Dam on the US-Mexico border and into the delta, to occasionally reach all the way down the river channel to the Sea of Cortez.

This is not to say there is no water in the region. Water diverted from the river at Morelos Dam, along with pumped groundwater, irrigates at least 450,000 acres, an area slightly larger than the Imperial Irrigation District on the northern side of the border.[5] Like their neighbors in the United States, the government and people of Mexico decided to use their share of the Colorado River to fuel a modern agricultural and urban economy. Like the Imperial Valley, the resulting landscape has been turned into incredibly productive farmland as a result, with wheat, cotton, and alfalfa the dominant crops on the Mexican side of the border. As in the United States, agriculture uses most of the water—91 percent by one estimate.[6] There is usually no water left for the river itself.

Scientific data is scant, but locals say beavers were nearly completely gone from the region during the dry times that came with the closure of Glen Canyon Dam and the diversion of the river's entire flow. But on the few occasions that the delta flooded, the beavers would reappear, perhaps following the flow down from refuges upstream.[7]

And so it was again, in the spring of 2014. A small flow of excess agricultural water flowed past willows through a human-built environmental restoration site. As soon as the water arrived, delivered through irrigation canals in an early phase of the river restoration efforts, beavers materialized out of the ecological mists, damming the little channel. They had found their way back.

By mid-summer, the beavers did not have to depend on the excess farm water. For a brief few weeks, Colorado River water flowing back down the river's main channel reached the environmental restoration site and a real river brought water to the delta's beavers.

It is important not to focus too narrowly on the spring 2014 "pulse flow" when the gates of the Morelos Dam were lifted. The environmental water was part of a historic amendment to the 1944 US-Mexico treaty, which created new rules for managing reservoir storage, shortages, and surpluses on the shared river. Even without the pulse flow, it would stand as a huge breakthrough in Colorado River management, a major step toward bringing the United States and Mexico together into a unified regime for the overstressed Colorado River. But the environmental piece of the agreement carried with it enormous symbolic and substantive value. As we stood at the foot of Morelos Dam in March 2014 watching the first water flow south, environmental attorney Peter Culp, who had helped negotiate the deal, said it was the first time two nations had used water to provide environmental flows across an international boundary.[8]

"Wasted Water"

Indeed, environmental flows had long been at the bottom of the priority list. If anything, water managers' efforts to use every last drop over the first decade of the twenty-first century had been making things worse for the delta. No project better illustrated the tension between environmental and water-management goals than the Drop 2 reservoir, off Interstate 8 in the Imperial Valley of southeastern California. As a water-conservation measure, Drop 2 was brilliant. Until the reservoir's construction was completed in 2010, a rainstorm in the Lower Colorado River Basin desert could perversely result in "wasted" water. Farmers would order water from Lake Mead, but if it rained during the three days it took the water to travel from the dam to their field, storm water from desert arroyos would add to the river's flows. At the same time, farmers' demand would drop because of the rain on their fields, resulting in extra water in the Colorado's channel. The result was water slipped past Imperial Dam and into Mexico unused, "wasted" in the eyes of US water users.[9] The amount thus "wasted" was small.[10] But it represented some of the only water that reached the otherwise dry riverbed south of Morelos Dam.

For cities desperate for every last drop, the potential for savings could not be ignored. The new Drop 2 reservoir (later renamed Warren H.

Brock Reservoir after a pioneering Imperial Valley farmer),[11] was built to capture the unused water. The next time the Imperial Irrigation District needed water, they could take some out of the new reservoir, leaving a like amount in Lake Mead.

The big urban water agencies, primarily Las Vegas, paid the reservoir's $200 million cost in return for credits for the saved water.[12] It was a clever water-management innovation, but it meant the end of some of the meager accidental environmental flows that otherwise wound up headed toward the delta.

For residents of Mexico as well as US environmentalists trying to figure out how to put water back into the delta for its own sake, this seemed like a step in the wrong direction. But at the same time as the Drop 2 project was moving forward, quiet discussions among water agencies and between the United States and Mexico were beginning to gain momentum. In the summer of 2007, US interior secretary Dirk Kempthorne and Arturo Sarukhan, Mexico's ambassador to the United States, met in Washington, DC, for what one of the participants described as "a very frank, informed, and thoughtful dialogue on the need for better Colorado River relationships between the two nations." Those discussions led to the issuance of a joint statement pledging to do something about the delta's environment in the context of broader US-Mexican negotiations to sort out unresolved differences over the Colorado River. The statement sketched out a broad agenda, from climate change to the possibility of seawater desalination to augment supplies. But it also made the most explicit acknowledgment to date that the delta mattered, highlighting "environmental priorities, including Colorado River Delta habitat protection and enhancement" as one of the key priorities for the talks to come.[13]

The statement promised "expedite(d) discussions" in a matter of weeks. The discussions began. A deal took longer. But the process was under way.

The United States and Mexico Start Talking

The steps that led to the beavers' water flowing back down the Colorado River's channel are a case study in how Colorado River Basin problem solving can succeed—including the part about how it will not be easy.

Philip Fradkin's 1981 classic *A River No More* framed much of the debate that followed, leaving a lasting impression of a delta lost for good. Fradkin called his closing chapter "Death in the Desert," describing a sad search through a wasteland of mudflats for the river's terminus.[14] Elsewhere the problems of North America's great estuaries—the Chesapeake Bay, the Everglades, the Sacramento–San Joaquin delta—triggered societal handwringing. As upstream users diverted their water and polluted what was left, the fate of these wetlands remained uncertain, but at least there was a societal conversation about them, with arguments over Endangered Species Act obligations and federal funding to try to fix the problems. But each of those estuaries lay entirely within the United States. In western North America, the convenience of an international border allowed us to largely ignore the Colorado River Delta, using the river's water on both sides of the border while ignoring the environmental and cultural consequences downstream.

Yet when the Colorado's great reservoirs "filled and spilled" in the late 1990s, scientists were amazed at the results. With a relatively small amount of water briefly returned to the river's main channel, it took little time for cottonwoods and willows along the old river's banks to begin popping back. The hopelessness of Fradkin's powerful account was unwarranted, scientists and environmentalists realized. While tearing down the big dams and replumbing the societies in both countries that depended on them was obviously unrealistic, the events of the 1990s made it clear that the amount of water needed to bring some semblance of life back to the delta river channel seemed within reach.

Environmentalists and scientists like the University of Arizona's Karl Flessa committed themselves to the goal of moving beyond the occasional accidental flows when the upstream reservoirs were full to creating a formal program to routinely put small amounts of water back in the delta's river. Even a modest pulse flow released from Morelos Dam down the main river channel, they argued, could bring the ecosystem back to life, albeit modestly. But the political entanglements of two nations, a border, and unsettled legal questions about who was entitled to how much water left the deal just out of reach.

On paper, the idea won early diplomatic support. In 2000, the two nations signed a formal agreement known as "Minute 306" to begin a formal binational conversation about putting water back in the delta. (Called a "minute" after the arcane diplomatic process that generated it—the agreement reads like the "minutes" of a meeting between the diplomats of the two nations—the deal exploited a much-used process that allows negotiated tweaks to the 1944 treaty between the two nations. They are technically not "amendments" to the treaty, but have the same effect.)

The agreement talked of "joint studies" and the need to "establish a framework for cooperation."[15] But it was toothless—just an agreement to talk, not to act. Crucially, the arena of action seemed to be all wrong. The deal had been signed by diplomats, but the paper document seemed far removed from the real world of moving water around the Colorado Basin. In practice, among water managers on both sides of the border the suggestion of somehow finding enough water in the system to send a pulse of water down into the delta was a hard problem.[16] Moving beyond the accident of filled reservoirs and excess water to intentional restoration seemed in the years that followed to be beyond the Colorado River management system's capabilities. Human demands upstream were just too great. How could water users possibly give up a share for the beavers' sake?

By the mid-'00s, people working on delta restoration began to realize that any agreement to put water back in the Colorado River's channel had to be about far more than just putting water back in the Colorado River's channel. "It started becoming clear that we needed to look at what we were doing in the delta in the context of bigger water-management decisions," explained Jennifer Pitt of the Environmental Defense Fund.[17]

Pitt, whom *High Country News*'s Matt Jenkins described as "a woman with a seemingly pathological drive to decipher the river's byzantine operational mechanics,"[18] had worked for years to learn the details of river management. Instead of disdaining the world of the "water buffalos," she embraced it, learning the complex computer simulations the US Bureau of Reclamation and other water agencies used to track the Colorado's stocks and flows.[19] Any solution to the environmental part of the problem, Pitt and her colleagues came to realize, would have to be linked to the costs and benefits of problem-solving efforts across the Lower Colorado River Basin.

The discussions seemed stalled until the ground shook, literally, on Easter Sunday in the spring of 2010. Two people died and at least 233 people were injured in the magnitude 7.2 quake, centered in the farming region of the Mexicali Valley on the western flank of the Colorado River Delta. The earthquake damaged the irrigation system used to deliver water to some 80,000 acres of farmland.[20] It was a striking demonstration of the vulnerability of Mexico's aging and fragile regional water-distribution system, according to an analysis by Vicente Sanchez and Alfonso A. Cortez-Lara, researchers at El Colegio de la Frontera Norte.[21]

The system's seismic vulnerability was certainly a problem, but also an opportunity, and the water-management institutions quickly kicked into gear. With no official deal in place, US and Mexican water managers immediately agreed at an informal level to reduce Mexican deliveries, storing the unused water in Lake Mead. For the first time in history, Mexican water was being quietly stored behind a US dam.

Then, in what passes for haste in the normally ponderous world of international diplomacy, representatives of the two nations negotiated an agreement that implemented on a temporary basis an idea that had been kicking around for several years without a tool for its implementation. A new accounting category was created on the US water ledgers for what came to be called "earthquake water"—unused Mexican water now stored in Lake Mead.

Interior Secretary Ken Salazar wasn't shy about sketching out the importance of a deal. It was, he said in a Mexico City announcement, "a remarkable achievement from a humanitarian perspective, but it also lays important groundwork for a much-needed comprehensive water agreement with Mexico on how we manage the Colorado River."[22] The earthquake had been turned from problem to opportunity, Mexican water was sitting behind Hoover Dam, and the diplomatic logjam was broken.

The Network in Action

Much of the negotiations that followed hinged on the development of a shared understanding of a seemingly straightforward problem: how to account for the water moving through the Colorado River System from one nation to the other. But to deal with the issue required doing something that had previously been beyond reach: extending "the network" of US water managers into Mexico.

One of the network's key tools is a sophisticated computer model of the basin known as the Colorado River Simulation System. The Bureau of Reclamation supported an entire research team based at the University of Colorado in Boulder to develop and maintain it, and its calculations are used to make decisions about how much, where, and when to release water.

Mexico's water agency, CONAGUA, operated an entirely different computer model of its own design, running on different software platforms, based on different assumptions about system operations, using different underlying data sets, and therefore yielding different results.

The disconnect illustrated one of the crucial insights from Elinor Ostrom's early research on California groundwater-basin management a half century before: a shared understanding of the resource, including the detailed math needed to measure the water's flow, is critical to any agreement. So one of the most important steps was entirely technical in nature—a series of exchanges that ultimately allowed the Mexican technicians to understand how the CRSS model works and to operate it for themselves. It let negotiations proceed, a group of participants in the talks wrote, "on a common factual footing."[23]

That shared understanding helped, but language still posed difficulties. In the end, a small negotiating group from both countries holed up in a Tijuana hotel with side-by-side English and Spanish versions of the bilingual agreement in a marathon session in late 2012.[24]

Consensus on the numbers only got them so far; there was still fundamental disagreement over the final terms of the deal. It involved the water itself that would be used for the pulse flow. US officials insisted that that no American water would be sent to the delta. Instead, they offered to pay for water-system improvements in Mexico to free up water, with a part of the conserved water dedicated to the environment. But this made Mexican officials uneasy because of the perception back home that the rich gringos north of the border were unwilling to contribute the one resource that really mattered.

In the end, a reading of the final deal, dubbed "Minute 319," shows the two sides papering over the problem with an uncomfortable agreement that left the tough stuff for later. In the short run, US water agencies would make modest contributions ($21 million) for water-saving system improvements in Mexico. Mexico retained title to nearly all the saved water, and Mexico would in return use some of its conserved "earthquake water" for the pulse flow. The US municipal water agencies, contributing a significant share of the money for Mexican water-conservation measures, got a share of the saved water as part of a complex water-accounting swap. US environmental groups would contribute money to buy up additional water rights in order to provide water to maintain environmental restoration sites in Mexico. That allowed the modest first experiment in Colorado River Delta restoration to succeed.

Along with the environmental benefits, two other major pieces formed the deal's three-legged stool. The second leg was clarification about how shortages and surpluses on the river would affect Mexico's deliveries. Under the deal, Mexico shares modestly in shortage if Lake Mead drops below elevation 1,075, the level at which Arizona and Nevada begin to take shortages. It also allows Mexico to share in surpluses if and when there is extra water in the river. The third leg was operational—the ability for Mexico to continue to store unused apportionment in Lake Mead.

But the big stuff, where the real needs and opportunities lie, was left for later. The deal talks about $1.7 billion worth of water-saving improvements that are possible in Mexico, to be jointly funded by water agencies on both sides of the border, "that could generate sufficient volumes of water to benefit both countries." They range from adding reservoir capacity in Mexico similar to what was done with Drop 2 on the US side, to building a big new plant to desalinate ocean water on the Mexican coast in exchange for a reduced delivery obligation, freeing up that water for use in the United States. But unable to come to terms on the actual projects, the agreement embodies the promise that discussions will continue.[25]

Bringing Water Back to the Delta

In one sweeping agreement, Minute 319 formally brought two traditionally excluded groups, environmentalists and Mexicans, into the Colorado River's formal management framework. It also felled two myths. The first was that environmentalists and water managers could not compromise to achieve common goals. The second myth was deeply

The Colorado River returns to San Luis Río Colorado (© John Fleck).

embedded in our western narrative—the idea that the delta was dead, that environmental rejuvenation was impossible in a world of expanding populations and growing water demands.

Minute 319's crowning bauble, the pulse of water released from Morelos Dam into the dry channel of the Colorado, briefly brought the river back to life. On March 23, 2014, the day the first water of the pulse flow was to be released, I pulled my wife's Subaru into the dirt parking area across Baja Highway 2 from Morelos Dam. A fast-food breakfast lay on the passenger seat; the sun wasn't yet up and I was the first to arrive. Journalist Matt Jenkins showed up next, followed shortly by environmentalist Jennifer Pitt, University of Arizona scientist Karl Flessa, Yamilette Carillo-Guerrero of the Colorado River Water Trust, and many more of the people who had worked so long to make the pulse flow happen. The formal ceremonies, with dignitaries and speeches, were not scheduled until days later, but none of us wanted to miss the first pulse as the dam's gates were opened.

Beyond the success of a mutually beneficial deal, putting water back into the delta just plain felt good. John Entsminger, the Colorado-born water lawyer who now heads the Las Vegas metro area's main water utility, remembers the day as a young college student that a friend explained to him that the Colorado River didn't reach the sea. "It was appalling," he recalled.[26]

Bob Snow, a US government lawyer who helped negotiate the agreement, recalled an uncle from Ireland telling him, "You know, rivers really should reach the sea."[27]

Minute 319 has rightly been hailed as a milestone and a significant success story. But it also demonstrates the limitations of water-management deal making in a complex, evolving environment. The deal is short-term in nature, an experiment. While it was enthusiastically received in Mexico, support was not universal, with some agricultural interests viewing the water diverted to environmental purposes as "wasted."

The water had not even finished flowing before the water managers who arranged it began talking about a follow-on deal—the next agreement that they know will be needed in the ever-evolving management of the Colorado River.

CHAPTER 13

Conclusion

ARIZONA SENATOR JEFF FLAKE had sharp questions for Deputy Interior Secretary Mike Connor during a hearing in the fall of 2015. Arizona water managers had been working hard all year to reduce the state's water use, hoping to leave water in Lake Mead as a buffer against shortage.

"The number-one priority in Arizona is to make sure that when Arizona, or any other state, voluntarily contributes their water to the health of the Colorado system, the contributed water actually stays in the system and doesn't disappear along somebody else's canals," Flake said, looking up from his notes with a smirk. Flake asked Connor to pledge that the federal government would not simply reassign the water to some other user. "Without these assurances," Flake said, "obviously such preventative measures don't make sense." So far so good—the senator was trying to guard against the tragedy of the commons. But then Flake's argument jumped the rails: "It'd be like having a savings account and just being able to see your neighbor reach and grab money from it."[1]

Flake was in part playing to the hometown audience. The references to "somebody else's canals" and "some other user" were dog whistles

to Arizonans and to water-management insiders across the basin, sig-
naling his vigilance against the ever-present threat (real or imagined)
that California and the federal government might be in cahoots to steal
Arizona's water.

But beyond Arizona's historic political paranoia, Flake's question
highlights the core problem with current management of Colorado
River water. As we have seen again and again—in farm districts like
Yuma and Imperial, and in urban areas like Albuquerque, Phoenix,
Las Vegas, and Southern California—the communities that depend on
the Colorado are learning to thrive on less water. But how to manage
"saved" water remains the central dilemma.

The problem is the clumsy metaphor itself—the idea that Arizona,
in conserving water in 2015, had been contributing to its own personal
Colorado River water savings account. A much more apt metaphor
would be that, in leaving water in Lake Mead, Arizona is simply trying
to prevent the account from being overdrawn. Given that, overall, water
users are demanding more water than nature is providing, sooner or
later the water that is conserved will have to be water that no one will
ever get back.

Albuquerque

My home town of Albuquerque, New Mexico, is a perfect case study of
the problem. It is no exaggeration to say that we have become very, very
good at conserving water. The question of what to do with the water
we've saved illustrates, in microcosm, the problems facing the Colorado
River going forward.

In the early 1970s, the US Bureau of Reclamation opened the gates
on the San Juan–Chama Project, which diverts Colorado River water
outside the river's hydrologic basin to be used in central New Mexico,
with Albuquerque getting the largest share. The project was intended to
shore up supplies on the Rio Grande, a relatively small river even by the

standards of the arid West, which by the 1950s was over-allocated.[2] It also marked the culmination of an effort to bring more water to a rapidly growing and thirsty city. A burst of Cold War growth, fueled by one of the United States' major nuclear-weapons research centers, made the acquisition of water essential, the state's leadership told Congress as it was considering legislation to fund the San Juan–Chama Project. Even with the new water, the region was believed to be at risk of outstripping its supplies.[3]

In the 1980s and 1990s, it looked as though that risk was becoming real. Despite the imported Colorado River water, the growing metropolitan area continued to lean heavily on its aquifer, and groundwater levels beneath some parts of the city had dropped more than a hundred feet.[4]

In response, the community in the mid-1990s launched a massive water-conservation effort that over two decades cut per capita water use nearly in half. While population grew modestly in the years that followed, water use kept dropping. The community spent some $500 million on a new water-distribution system that allowed more efficient use of the imported San Juan–Chama water, and by 2015 the aquifer was rising across the big groundwater basin that underlies the Albuquerque metro area.

It took me a long time to grasp that this was actually happening. As a newspaper reporter covering the region's water supply, I remained steeped in old narratives of crisis, of communities at risk of running out of water. Monthly, I asked for copies of the water utility's pumping reports, and I made a habit of routinely checking the US Geological Survey's groundwater monitoring well data, ready to pounce at the first sign of failure. But our water use keeps dropping and our aquifer keeps rising. Grudgingly at first, but then with increasing fascination, I began writing stories about what success looked like.

This has created an entirely different problem for the community's water managers. By the end of 2015, the Albuquerque Bernalillo

County Water Utility Authority estimated that the aquifer beneath the city held nearly a million acre-feet more water than it would have held absent the conservation efforts. While the size of that number is based on generous assumptions made by the agency, there is no question about which direction water management and policy in the New Mexico city are now headed. Between its imported San Juan–Chama water and its rights to local groundwater and surface water, Albuquerque has rights on paper to 32 percent more water than it is currently using. Instead of scrambling to find new water, the agency has shifted to finding ways of managing the surplus—expanding storage space in a big northern New Mexico reservoir and ramping up efforts to inject water into the aquifer for later use.[5]

But nowhere in the community's water-management dialogue is there any mention of what might seem like an obvious response to the current situation: maybe Albuquerque shouldn't take so much water from the over-allocated Colorado River Basin every year? Albuquerque's water managers could offer a rational response, of course. If they left San Juan–Chama Project water unused, to paraphrase Flake's question to Connor, wouldn't it just end up in someone else's canals, watering someone else's golf courses and lawns?

What to Do with Saved Water

Albuquerque is not unique.

Las Vegas, as we have seen, has demonstrated remarkable water-conservation success, using just 75 percent of its Colorado River allocation in 2014. But it still claims the full amount, stashing the surplus in groundwater storage projects scattered around the West. By the end of 2014, southern Nevada's municipal water managers had more than a million acre-feet of water in its water banks. The result is that Lake Mead drops just as much as if Las Vegas had not conserved a drop.

Consider Phoenix, which has important similarities to Las Vegas and

Albuquerque, as well as important differences. In 2014, Phoenix used just 70 percent of its 186,557 acre-foot Colorado River water-rights entitlement. But because of the "use it or lose it" rules under which the Arizona water rights are administered, Phoenix had no way to simply leave that water in Lake Mead for the good of the overall Colorado River system. Instead, any of Phoenix's allocation that went unused simply reverted to other users in Arizona. Where Albuquerque and Las Vegas have an incentive to conserve and hoard the water for the cities' own growth and future use, Phoenix has little incentive to conserve water at all.[6]

The agricultural districts of Yuma are yet another example of the problem. As we have seen, the farmers there, in shifting to lucrative winter vegetable production, have reduced their water use by 30 percent since the 1970s. But their reduction in water use did nothing to reduce the overall pressure on the over-allocated Colorado River water system.

Lettuce fields outside Yuma, Arizona (© John Fleck).

Instead, it was simply shifted to other users in Arizona, with the resulting pressure continuing to drop Lake Mead.

Given our current rules, all these communities are behaving as rational actors, making choices in their own best interests. The latest long-range planning data being developed by the Albuquerque Bernalillo County Water Utility Authority clearly shows that my community could absorb a substantial reduction in its allocation of Colorado River water. But until the rules create a mandate or incentive to conserve, it would be crazy for Albuquerque to voluntarily relinquish the water without some assurance of the sort sought by Senator Flake.

Defining the "Crisis"

So what might this collective action look like? In the end, we need an honest reckoning with the basic problem: there is not enough water for everyone to do everything they want with it, or to use every drop to which they feel legally entitled.

It is frequently said that the Colorado River Basin is in "crisis," but the detailed nature of that "crisis" is rarely spelled out. It is often framed in terms of the relationship between the river's supply and growing demand, and illustrated with pictures of the white "bathtub ring" circling the increasingly empty Lake Mead. But when demand rises above supply, and reserves in the basin's great reservoirs run out, the basin's water users cannot consume negative water. Someone will have to use less, or stop using altogether. Who is that? In defining the problem, then, we sooner or later have to get specific. When shortages occur, who takes the hit?

This is the point at which the river's operating rules, and most importantly, the ambiguity about how these rules will be implemented in times of scarcity, become critical. It is also the place at which we can dimly make out the shape of long-term solutions. Without those solutions, the current rules and the physical reality of the system suggest five big risks.

The most immediate risk is Las Vegas. As we have seen, Las Vegas has demonstrated the ability to live within its means, dropping its usage over the first years of the twenty-first century so that it now consumes substantially less than its 300,000 acre-feet per year allocation. But it continues to face a physical risk. As Lake Mead drops, it becomes harder and harder for Las Vegas to get water from its intake pipes, with a clear risk that the reservoir's level might drop so low that Las Vegas could have a legal entitlement to water that it has no physical way to get to the city.

Las Vegas is taking steps to deal with the problem, building a new intake that is deeper in the reservoir, and a new pumping plant to handle the deeper water. The first phase, the deeper pipe to take the water, opened in 2015. The completion of the pumping plant, by 2020, will eliminate the largest risk in the entire basin, which is a city of 2 million people going dry because its intake pipes are above the water line, sucking air. With a new intake deep within Lake Mead, Las Vegas's water supply will ironically be transformed from one of the most vulnerable to one of the most secure in the Lower Basin.

The second risk is to Arizona. As Lake Mead continues to drop, and supplies become increasingly constrained, the operating rules require that the major burden falls on Arizona. In return for the Central Arizona Project's construction, California extracted a legal requirement (enshrined in federal law) that all of Central Arizona's Colorado River allocation, all 1.5 million acre-feet per year that flow through the big canal to Phoenix and Tucson must be cut off completely before California loses a single drop. Arizona has long viewed subsidized agriculture as a buffer, and it has been banking surplus groundwater as a hedge against such an inevitability. This, too, will be tested.

The third risk is to the states of the Upper Colorado River Basin. Most legal scholars agree that if (when?) the total flow in the river drops because of climate change, the rules enshrined in the Colorado River Compact require the Upper Basin to continue sending 7.5 million acre-feet downstream past Lee's Ferry each year. If climate change push comes to shove and there is not enough water to meet that requirement and also supply all the current water use needs in Wyoming, Colorado, Utah, and New Mexico, those Upper Basin water uses must be cut to meet the Lee's Ferry delivery requirement.[7]

The fourth risk is the long-festering problem of Native American communities' rights to water. Unresolved, this uncertainty leaves these communities without the water they need to prosper, and it also leaves a cloud of uncertainty over other water users.

The fifth risk always remains the delta and, more broadly, the environment. The operating rules have long left environmental flows entirely out of the picture, and it is only slowly and at great pain and expense that water has been carved out to bring the environment back.[8]

In a remarkable talk in the summer of 2013, Colorado water scholar Brad Udall argued that when we debate the legal fine points about who is entitled under the law to how much water, we're debating the wrong things. Udall's father, Arizona congressman Morris "Mo" Udall, helped build the modern Colorado River plumbing system, and the younger Udall has now become one of the leading public intellectuals in working through the set of problems that we must now cope with in the twenty-first century. Speaking at the University of Colorado School of Law to an audience of lawyers, law students, and legal scholars, Udall argued that all these arcane, technical legal arguments over the future of the Law of the River are beside the point.

Udall distinguished between the "reality of the public" and the "reality of the water community," describing a world in which regular folks have no clue about things water professionals obsess over—things like Article III(d) and the Colorado River Basin Storage Project Act and the "doctrine of prior appropriation." Udall's "public" is a community of people who just want to turn on the tap and have the water come out. But they have some basic notions of fairness and good sense, imagining that the policies underlying our attempt to supply that water will consider questions of equity, sound economics, and the environment. Water-management actions that violate those notions will, to quote Udall, "violate the public's sense of 'rightness.'"

This means several things in terms of practical Colorado River Basin management:

It means that, whatever the Law of the River says, as a practical matter we're not going to let the level of Lake Mead drop below 1,000 feet above sea level, the surface elevation at which Las Vegas can no longer get water from the reservoir. Udall quoted Mike King, head of the Colorado Department of Natural Resources: "I don't care what you think about the Law of the River, we are not going to dry up a city of 2 million people."

It means that the public will not tolerate a situation in which Phoenix loses all its Central Arizona Project water while California continues to take a full allotment.

And it means, Udall argued, that the states of the Upper Basin will not be left to take the full brunt of the river's shrinkage because of climate change while the Lower Basin gets its full Compact share.[9]

The events of the last two decades—the 2001 negotiations to wean California of its dependence on surplus water, the 2007 deal to share shortages, the US-Mexico agreement embodied in Minute 319—provide an indication of what the path forward might look like. They do not suggest that the Law of the River's imperfections and ambiguities are irrelevant. Instead, they suggest that "the network" has come to the shared conclusion that arguing over legal interpretation is the wrong path. Those ambiguities and imperfections are flaws to be corrected through collective action and agreement rather than winner-take-all legal battles.

What we need is a negotiated solution that avoids narrow interpretations of the Law of the River, that reduces allocations broadly, enables trading among water users, and capitalizes on the reality that, as we have seen, communities in the Colorado River Basin have a remarkable ability to preserve their ways of life when supplies run short.

Connor's Response

Connor's response to Flake's questioning shows what solutions to the Colorado River Basin's problems might look like. As the two senior officials sparred in public, speaking in the coded, theatrical language of a congressional hearing, behind the scenes negotiators for the states and water agencies had been scrambling to come up with a deal that would slow the decline in Lake Mead and share the pain of shortage. A series of small, interim deals had been approved under which water users would be compensated to forgo water that would remain in Lake Mead, unallocated, as what the water managers had come to call "system water"— water that was in a sense no one's water, but also everyone's. But those "system water" agreements had generated only small amounts of water, and the negotiators seemed powerless to halt the reservoir's continuing decline. During the fall hearing of 2015, Lake Mead was at its lowest October level since 1936, when the big reservoir was first being filled.

Connor's response was that the federal government would not step in unilaterally. As in the 2001 agreement that finally curtailed California's overuse of Colorado River water, and again in 2007 when the states came together to approve their first shortage-sharing agreement, it was up to the water users to find a solution. The federal government would play an important role in both the negotiation and the implementation, but it would not impose an answer. "The states' agreement," Connor said, "has been the model for us to operate."

Eighteen months earlier, Connor and I were riding down a levee road along the banks of the last stretch of the Colorado River on the US-Mexico border, watching water from the Minute 319 pulse flow fill a stretch of the riverbed that was normally dry. It was the day after the triumphant international pulse flow celebration staged at Morelos Dam, and I had tagged along with a delegation of senior water managers and federal scientists who had ditched diplomatic formalities and press handlers to get a firsthand look at the water.

Connor and Assistant Secretary of the Interior Anne Castle acted like eager tourists, pulling out their cell phones to take pictures at every stop. And as we drove, they reflected on the process that had gotten everyone to this remarkable moment when the representatives of nine states and two nations, plus a larger coalition of water agencies, environmentalists, and community groups, had come together to solve what had seemed an insoluble water-sharing problem—putting water back in the Colorado River Delta.

Minute 319 was the logical extension, Connor explained to me, of a process in which water users stepped away from Marc Reisner's old refrain of the Colorado as the world's "most litigated" river to the approach of the past two decades—of negotiated agreements to deal with the shortcomings of the current water-management rules. "The Law of the River," Connor told me, "is pretty all-encompassing, but it still has room for interpretation."

It was a jubilant day, but as we turned from the river to ferry Connor back to the Yuma airport for his flight home, the talk turned to the serious next steps. At that time, the Bureau of Reclamation's internal modeling efforts showed the potential for drought and climate-change trouble ahead. As Lake Mead drops, rules kick in that require water users in Nevada, Arizona, and Mexico to remove less water from the system each year. But those reductions are modest, and Connor told me that the Bureau's worst-case modeling showed that even with the agreed-upon reductions, Lake Mead could quickly drop past a point of no return, to levels at which the current rules would be no help in determining who was entitled to how much.

The solution is, in a sense, straightforward. Everyone in the Colorado

River Basin has to use less water. It's possible to apply a simple arithmetic wave of the arm and say, for example, that we could bring the system into balance if everyone used 20 percent less water than they are consuming today. We know from experience, from Yuma to Las Vegas to Albuquerque, that such reductions are possible, that water-using communities are capable of surviving and even thriving with substantially less water than they use today. But no one will voluntarily take such a step without changes in the rules governing basin water use as a whole to ensure that everyone else shares the reductions as well—that any pain is truly shared. We need new rules. Absent that, we simply end up with a tragedy of the commons.

Where do those rules come from?

The answer, Connor said, was not some grand solution imposed from on high by the US secretary of interior and other agencies of the federal government, but a process of continued negotiation among the federal government and the states to come up with better rules to cope with shortfalls and decide who must use less water. It is the kind of process that Ostrom so insightfully observed in the formation of West Basin water-management tools, and that we have seen time and again as communities come together to solve common-pool resource-management problems large and small. Connor's answer as we drove back through the farmland of Yuma County, past lettuce fields nearing the end of their harvest season, was, in concept, identical to what he told Flake in front of reporters and television cameras at the autumn 2015 Senate hearing: "It starts with the seven basin states and their ongoing dialogue."[10]

Beyond the Myths

The biggest barrier to that dialogue is believing our own myths. We have seen, time after time, the myths of conflict, water flowing toward money, and—most importantly—crisis, all dispelled by communities who were able to compromise and conserve. Yet the mythology has real staying power.

In January 1960, the *Los Angeles Times* featured a frightening front page headline: "Southland's Water Safety Margin Placed at 10 Years." Groundwater basins would go dry, and the region's rapidly rising population would soon outstrip its supplies of imported water, the newspaper breathlessly warned.[11] It was neither the first nor the last headline in a US newspaper to play on the fear that the West is running out of water. Recent years, especially during the drought in the twenty-first century, saw a rash of similar articles: "Scarce Water and the Death of California Farms," "The Dust Bowl Returns," "A 'Megadrought' Will Grip US in the Coming Decades," "Colorado River Water Supply to Fall Short of Demand."

These dire stories fuel fears that the only way to save yourself is to fight your neighbors for every last drop. But if you are able to sidestep the crisis narrative and recognize that your community can thrive with less water, then the fight with your neighbors seems less necessary and the risks of water wars and a crash diminish. It's time to stop fighting over water and turn our attention to another adage Mark Twain likely never said: "The secret of getting ahead is getting started." When we start talking, we can learn to share our beloved but dwindling Colorado River in a changing world.

Notes

Note: The following notes are presented in an abbreviated fashion. Full citations for each publicly available source can be found in the bibliography.

Chapter 1 Notes

1. Sykes, *The Colorado Delta*, 161.
2. MacDougal, "The Delta of the Rio Colorado," 10.
3. Grant, "Tapping the Past for California's Water Future."
4. See, for example: Deverell and Sitton, "Forget It, Jake," 3; Erie, *Beyond Chinatown*, 4.
5. Powell, *Dead Pool*, 240.
6. Sabo et al., "Reclaiming Freshwater Sustainability in the Cadillac Desert," 21263–4.
7. Reisner, *Cadillac Desert*, 501.
8. US Bureau of Reclamation, "Colorado River Basin Water Supply and Demand Study, SR-26," C-9; US Bureau of Economic Analysis Regional Economic Accounts, "Annual Gross Domestic Product (GDP) By State"; US Department of Agriculture, "2012 Census of Agriculture," 245–52.
9. Fleck, "State's Future Banks on Colorado River."

Chapter 2 Notes

1. US Department of Agriculture, "2012 Census of Agriculture, New Mexico State and County Data," 226–28.
2. Fleck, "Farming Against the Odds on the Rio Grande."
3. US Geological Survey, "Largest Rivers in the United States," 2.
4. Kuhn, "The Colorado River: The Story of a Quest for Certainty on a Diminishing River," 22.
5. Minutes of the Twelfth Meeting of the Colorado River Commission, November 12, 1922, 73.
6. Woodhouse, Gray, and Meko, "Updated Streamflow Reconstructions for the Upper Colorado River Basin," 4.
7. Hundley, *Water and the West*, 274.
8. US Bureau of Reclamation, "Colorado River Basin Water Supply and Demand Study, SR-26," SR-4; US Bureau of Reclamation, "Colorado River Basin Natural Flow and Salt Data."
9. Nash and Gleick, *Colorado River Basin and Climatic Change*, ix; US Bureau of Reclamation Colorado River Basin Supply and Use Data, 2015 (dataset provided by the US Bureau of Reclamation, September 22, 2015); Vano et al., "Understanding Uncertainties in Future Colorado River Streamflow," 59; US Bureau of Reclamation, "Lower Colorado River Operations, Lake Mead at Hoover Dam."
10. Cohen, Christian-Smith, and Berggren, *Water to Supply the Land*, 7.
11. Hays, *Conservation and the Gospel of Efficiency*, 11.
12. US Bureau of Reclamation, "Colorado River Basin Water Supply and Demand Study, SR-26," C-9.
13. US Bureau of Reclamation, "Moving Forward: Phase 1 Report," 4–7.
14. US Bureau of Reclamation, "Moving Forward: Phase 1 Report," 4–6.
15. Russelle, "After an 8,000-year Journey, the 'Queen of Forages' Stands Poised to Enjoy Renewed Popularity," 252.
16. Putnam et al., "The Importance of Western Alfalfa Production," 2.
17. US Department of Agriculture National Agricultural Statistics Service, "CropScape Cropland Data Layer."
18. Medellín-Azuara, Lund, and Howitt, "Jobs per Drop Irrigating California Crops."
19. Cohen et al., "Water to Supply the Land," 61.

20. Hundley, *The Great Thirst*, 469.

21. Nelson, "Increase in Crop Acreages and Property Values in Imperial County," 2.

22. Sauder, *The Yuma Reclamation Project*, 55.

23. US Bureau of the Census, "Sixteenth Census of the United States: 1940," 409.

24. US Department of Agriculture, "Census of Agriculture, 2012," Arizona 276; US Bureau of the Census, "Sixteenth Census of the United States: 1940," 399, 409.

25. Imperial County Agricultural Commission, "Imperial County Agricultural Crop and Livestock Report," 2000, 3; 2014, 5.

26. US Department of Agriculture, "Fresh Fruit and Vegetable Shipments, 2014," 24–25.

27. US Department of Agriculture, "Census of Agriculture, 2012," Arizona, 226–28; California 239–47; Colorado 227–36; New Mexico 225–30; Nevada 212–16; Utah 224–28; Wyoming 223–27.

28. Yuma County Agricultural Water Coalition, "A Case Study in Efficiency," 30–35.

29. US Bureau of Economic Analysis, "Regional Economic Accounts, Farm Income and Expenses" (table CA45).

30. Bureau of the Census, "Census of Agriculture, 1974," Arizona 11–12; US Department of Agriculture, "Census of Agriculture, 2012," 2014, Arizona 276; US Bureau of Reclamation, "Compilation of Records in Accordance with Article V of the Decree of the Supreme Court in *Arizona v. California*," 8–9; US Bureau of Reclamation, "Colorado River Accounting and Water Use Report—Arizona, California, and Nevada," 2015, 14–16.

31. US Bureau of Economic Analysis, "Regional Economic Accounts, Farm Income and Expenses" (table CA45).

32. Cohen, Christian-Smith, and Berggren, *Water to Supply the Land*, 69.

33. Khaled Bali, author interview, July 7, 2014.

34. Fleck, "When Water Supplies Ebb, Users Go with the Flow."

35. US Department of Agriculture, "Crop Production—2014 Summary," 32–33.

36. California Department of Food and Agriculture, "Dairy Production Data"; US Department of Agriculture, "Milk Production."

37. Dan Putnam, author interview, September 15, 2015.

Chapter 3 Notes

1. Woestendiek, "In Middle of Desert, Las Vegas Builds Lake."
2. Moehring, *Resort City in the Sunbelt*, 46.
3. *Las Vegas Sun*, "How Much Water Evaporates from the Bellagio Fountains?"; Bali, Division of Agriculture and Natural Resources, "Imperial Alfalfa Irrigation Requirement."
4. Author's estimate based on reported hotel rooms, occupancy rates, and visitor spending reported by the Nevada Gaming Control Board (see the "Nevada Gaming Abstract 2014").
5. Nevada Gaming Control Board, "Nevada Gaming Abstract 2014," 1–10.
6. US Department of Agriculture, "2012 Census of Agriculture—Imperial County California Profile," 1.
7. Davis, *Dead Cities*, 101. See also Harrison, "Water Use and Natural Limits in the Las Vegas Valley," which critiques Davis and other Las Vegas critics.
8. Gleick and Palaniappan, "Peak Water Limits to Freshwater Withdrawal and Use," 11161.
9. Moehring and Green, *Las Vegas*, 2.
10. Moehring, *Resort City in the Sunbelt*, 4.
11. Dumke, "Mission Station to Mining Town: Early Las Vegas," 263; Vegas Artesian Water Syndicate, prospectus.
12. *Las Vegas Age*, "Action of 7 States Means Millions to Las Vegas."
13. Moehring, *Resort City in the Sunbelt*, 212.
14. Southern Nevada Water Authority, "Water Resource Plan 2009," 2.
15. *Arizona v. California*, 373 U.S. 546 (1963), 583.
16. Moehring, *Resort City in the Sunbelt*, 212.
17. Ibid., 213.
18. Ibid.
19. Harrison, "Water Use and Natural Limits in the Las Vegas Valley," 39.
20. Ibid., 40.
21. Ibid., 44.
22. Ibid., 40.
23. US Bureau of Reclamation, "Compilation of Records in Accordance with

Article V of the Decree of the Supreme Court of the United States in *Arizona v. California*," 1987, 26; US Bureau of Reclamation, "Compilation of Records in Accordance with Article V of the Decree of the Supreme Court of the United States in *Arizona v. California*," 1993, 18.

24. Southern Nevada Water Authority, "Water Resource Plan, 2009."
25. Harrison, "Water Use and Natural Limits in the Las Vegas Valley," 50.
26. Mulroy, "Beyond the Division," 105.
27. Kurtis Hyde, author interview, February 26, 2015.
28. William Hasencamp, presentation to Metropolitan Water District of Southern California Water Planning and Stewardship Committee, July 13, 2015; Brean, "Authority Approves Leasing Water to California."
29. Southern Nevada Water Authority, "Water Resource Plan 2009," 67; US Census Bureau, "2014 Population Estimates, QuickFacts."
30. John Entsminger, author interview, October 1, 2015.
31. Southern Nevada Water Authority, "Member Agency Estimated Population and Annual Water Usage 1994–2013" (data provided by Southern Nevada Water Authority, June 17, 2014); Albuquerque Bernalillo County Water Utility Authority, "New Mexico Office of State Engineering GPCD Calculator," 2014 (data provided by Albuquerque Bernalillo County Water Utility Authority, January 14, 2015).
32. Cahill and Lund, "Residential Water Conservation in Australia and California," 118.
33. Grant et al., "Adapting Urban Water Systems to a Changing Climate: Lessons from the Millennium Drought in Southeast Australia," 10728.
34. Las Vegas Valley Water Authority, "Consolidated Annual Financial Report," 2014, 71.
35. Glaeser, *Triumph of the City*, 132.
36. Lustgarten, "The 'Water Witch.'"
37. Southern Nevada Water Authority, "Water Conservation Plan 2014–2018," 22; Southern Nevada Water Authority, "Water Resource Plan 2015," 25; US Bureau of Reclamation, "Colorado River Accounting and Water Use Report, 2015," 22.
38. Los Angeles Department of Water and Power, "2005 Urban Water Management Plan," 1–9; California State Water Resources Control Board, "Urban Water Supplier Enforcement Statistics."

39. Albuquerque Bernalillo County Water Utility Authority, Municipal Pumping Report, 2014 (provided to author upon request); John Stomp, "Water Resources Management Strategy Update."

40. Southern Nevada Water Authority, "Water Resource Plan 2015," 39.

Chapter 4 Notes

1. Sid Wilson, author interview, May 25, 2015; Jennifer Pitt, author interview, November 12, 2010. Pitt worked for the Environmental Defense Fund during the events described in this book. In late 2015, she became Director of the Colorado River Project for the National Audubon Society.

2. Testimony of Jennifer Pitt, Environmental Defense Fund, US House Subcommittee on Water and Power, Oversight Hearing on Collaboration on the Colorado River, April 9, 2010.

3. Leavenworth, "Colorado Conversion?"

4. Carrillo-Guerrero, "From Accident to Management," 84; Jennifer Pitt, personal communication, September 18, 2015.

5. Glennon and Pitt, "Our Water Future Needs Creativity."

6. Ward, *Border Oasis*, 45.

7. Hundley, *Dividing the Waters*, 173.

8. Judkins and Larson, "The Yuma Desalting Plant," 410.

9. Yuma Desalting Plant / Cienega de Santa Clara Workgroup, "Balancing Water Needs on the Lower Colorado River," 1.

10. Ostrom, "Beyond Markets and States: Polycentric Governance of Complex Economic Systems," 641.

11. Gerlak, "Resistance and Reform: Transboundary Water Governance in the Colorado River Delta," 100–19.

Chapter 5 Notes

1. Cohen et al., "Municipal Deliveries of Colorado River Basin Water," 13–15.

2. US Geological Survey, "Estimated Use of Water in the United States."

3. Konikow, "Groundwater Depletion in the United States," 7.

4. Kathryn Sorensen, author interview, February 20, 2015.

5. Welsh, *How to Create a Water Crisis*, 14.

6. Martin and Young, "The Need for Additional Water in the Arid South-

west: An Economist's Dissent," 31.

7. Arizona Constitution, Article 22, Section 20.

8. Young, "The Arizona Water Controversy: An Economist's View," 3.

9. Kupel, *Fuel for Growth*.

10. Statement of Ralph Murphy, Hearings held at the Ingleside Inn, Phoenix, Arizona, on H.R. 6251 and H.R. 9826 before the House Committee on Irrigation and Reclamation, 69th Congress (1926).

11. St. Louis Federal Reserve, Resident Population in Arizona.

12. Statement of Fred T. Colter, Hearings on H.R. 2903 before the House Committee on Irrigation and and Reclamation, 68th Congress (1924).

13. "Dam Storm Thunders," *Los Angeles Times*, May 23, 1928; "Senators Battle Over Boulder Dam," *New York Times*, May 30, 1928.

14. Mann, *The Politics of Water in Arizona*, 83.

15. Beaver, *Images of America: Parker*, 86.

16. Reisner, *Cadillac Desert*, 258.

17. *Los Angeles Herald-Express*, November 14, 1934, in Beaver, *Images of America: Parker*, 85.

18. Summitt, *Contested Waters*, 42.

19. Hearings on Arizona water resources before a subcommittee of the Committee on Irrigation and Reclamation, United States Senate, 78th Congress, 2nd session, July 31 and August 1–4, 1944, 46.

20. Ibid., 9.

21. Ibid., 13.

22. August, *Vision in the Desert*, 175.

23. Motion for Leave to File Bill of Complaint and Bill of Complaint, *Arizona v. California*, No. [10] Original, 1952 Term (U.S.).

24. MacDonnell, "*Arizona v. California* Revisited," 370.

25. A bill to authorize, construct, operate, and maintain the Central Arizona Project, Arizona–New Mexico, and for other purposes, S. 1658, 88th Congress (1964).

26. *Yuma* (Arizona) *Sun*, June 9, 1963, 6.

27. August, *Vision in the Desert*, 187.

28. "Sen. Hayden Takes Step in Billion Dollar Dream," *Yuma Sun*.

29. Dozier and McCann, "CAP Priority to Colorado River Water," undated, 1.

30. Arizona Interstate Stream Commission, 27.

31. US Department of Agriculture, National Agricultural Statistics Service Quickstats.

32. Campbell, "Pinal County Agriculture," 13.

33. *Colorado River Basin Project Act*, Public Law 90-537, 90th Congress, S. 1004 (1968).

34. Connall, "A History of the Arizona Groundwater Management Act," 331.

35. Bureau of the Census, "1978 Census of Agriculture," Arizona 119; US Department of Agriculture, "2012 Census of Agriculture," Arizona 249.

36. US Geological Survey, "Estimated Use of Water in the United States in 1980," 36; US Geological Survey, "Estimated Use of Water in the United States in 2010," 9.

37. Kathleen Ferris, author interview, August 3, 2015; Konikow, "Groundwater Depletion in the United States (1900–2008)," 7.

38. US Geological Survey, "Estimated Use of Water in the United States," 1975 and 2010.

39. Loomis, "Ducey: Don't Punish Arizona for Its Water Conservation."

40. Davis, "Is California Trying to Take Our Water?"

Chapter 6 Notes

1. "Ground Water: The Perils to Its Purity," *Los Angeles Times*, December 12, 1989.

2. Ostrom, "Public Entrepreneurship," 13.

3. Western Regional Climate Center, Long Beach Daugherty Field, California.

4. Hundley, *The Great Thirst*, 88; Ostrom, "Public Entrepreneurship," 11.

5. Mendenhall, *Development of Underground Waters in the Central Coastal Plain Region of Southern California*, 10.

6. Ostrom, "Public Entrepreneurship," 15.

7. Hardin, "The Tragedy of the Commons," 1243–48.

8. Chibnik, *Anthropology, Economics, and Choice*, 161.

9. See, for example: Green, *Managing Water*, 64.

10. Ostrom, "A Long Polycentric Journey."

11. Ibid., 14.

12. Ostrom, "Why Do We Need To Protect Institutional Diversity?" 141.

13. Ostrom, "Public Entrepreneurship," 284.

14. "Warning Given on Salt Barrier," *Los Angeles Times*, July 26, 1946

15. Ostrom, "Public Entrepreneurship," 284.

16. Ibid., 30.

17. Ibid., 236.

18. Ibid., 238.

19. Ibid., 271.

20. Sarewitz, "How Science Makes Environmental Controversies Worse," 399.

21. Ostrom, "Public Entrepreneurship," 270.

22. Ostrom, *Governing the Commons*, 116.

23. Ostrom, "Public Entrepreneurship," 288.

24. Ibid., 294.

25. Blomquist, "Crafting Water Constitutions in California," 111.

26. Ostrom, *Governing the Commons*, 120.

27. Blomquist, *Dividing the Waters*, 302.

28. Ostrom, *Governing the Commons*, 126

29. Ibid.

30. "Hawthorne Mayor Hits 'Double Tax,'" *Los Angeles Times*, November 15, 1959.

31. Ostrom, Gardner, and Walker, *Rules, Games, and Common-Pool Resources*, 4; Water Replenishment District of Southern California, "Engineering and Survey Report," 13–16.

Chapter 7 Notes

1. US Bureau of Reclamation, "Accounting for Colorado River Water Use within the States of Arizona, California, and Nevada Calendar Year 2003," 15–16.

2. Hettena, "Interior Secretary Cuts California's Share of Colorado River Water."

3. Grace Napolitano, in testimony to the Subcommittee on Water and Power, Committee on Resources, H.R. Rep. No. 107-78 at 43 (2001).

4. Tyler, *Silver Fox of the Rockies*, 107.

5. The technical dividing line is a place called "Lee Ferry," a geographical artefact of the need to include outflow from the Paria River, located slightly downstream from the place more commonly known as "Lee's Ferry," where early Mormon settlers had operated a ferry. To further complicate matters,

the National Park Service, which now maintains the old site, calls it "Lees Ferry."

6. Lochhead, "Upper Basin Perspective on California's Claims to Water from the Colorado River, Part II," 328.

7. MacDonnell, "*Arizona v. California* Revisited," 363.

8. Larry Anderson in testimony to the Subcommittee on Water and Power, Committee on Resources, H.R. Rep. No. 107-78 at 11 (2001).

9. Western Regional Climate Center, "Historic Climate Division Data Summaries."

10. US Bureau of Reclamation, "Colorado River Basin Natural Flow Data."

11. Wood, "California Drought Springs New Limits on Developers," 9.

12. Lochhead, "An Upper Basin Perspective, Part II," 326.

13. Coates, "Colorado Offers Water to California."

14. Lochhead, "An Upper Basin Perspective, Part II," 328.

15. US Bureau of Reclamation, "Agreement Requesting Apportionment of California's Share of the Waters of the Colorado River Among the Applicants in the State" (commonly known as the "The California Seven Party Water Agreement of 1931"); Jerla, "An Analysis of Coordinated Operation of Lakes Powell and Mead under Lower Reservoir Conditions," 21.

16. Reisner and Bates, *Overtapped Oasis*, 149.

17. Glennon, *Unquenchable*, 258–71.

18. Lochhead, "An Upper Basin Perspective, Part II," 354.

19. Ibid., 356.

20. Ibid., 395.

21. Vollmann, *Imperial*, 956.

22. Minutes of Imperial Irrigation District Special Board Meeting, El Centro, CA, December 9, 2002.

23. Lochhead, "Upper Basin Perspective on California's Claims to Water from the Colorado River Part II," 396.

Chapter 8 Notes

1. US House Subcommittee on Water and Power, Committee on Resources, H.R. Rep. No. 107-78 at 43 (2001).

2. Phillip Pace, Minutes, Special Meeting of the Board of Directors of the Metropolitan Water District of Southern California, January 6, 2003, 258.

3. William Hasencamp, author interview, May 27, 2015.

4. Gottlieb and FitzSimmons, *Thirst for Growth*, 14.

5. Ibid., 15.

6. Hundley, *The Great Thirst*, 284.

7. California Department of Water Resources, *California State Water Project at a Glance*, 1.

8. Testimony of Ronald Gastelum, "Water: Is It the Oil of the 21st Century?" Hearing before the US House Committee on Transportation and Infra-structure, May 22, 2003.

9. Metropolitan Water District of Southern California, "Southern California's Integrated Water Resources Plan," 1-1.

10. Ibid., 3–35.

11. Metropolitan Water District of Southern California, *Annual Progress Report to the California State Legislature*, 19.

12. US Bureau of Reclamation, "Groundwater Banking Pilot Project of Central Valley Project Water from City of Tracy to Semitropic Water Storage District," 8.

13. Semitropic Water Storage District, Combined Financial Statements, 6.

14. Howitt and Hanak, "Incremental Water Market Development," 79.

15. William Hasencamp, author interview, April 30, 2015.

Chapter 9 Notes

1. Meister, "Sample Cost to Establish and Produce Wheat, Imperial County," ii; Imperial Irrigation District, "Monthly Crop Acreage Report," 1.

2. Zetland, "Will the People of Imperial Valley Jump or Get Pushed?"

3. Tina Shields, author interview, August 14, 2015.

4. Berman, "A Tale of Two Transfers."

5. "Get Ready for the Big Grab," *Imperial Valley Press*, February 25, 2007.

6. "Land, Water, Homes, Stability, and Progress," *Los Angeles Times*, August 9, 1903.

7. Clifford, *Overland Tales*, 299.

8. Meadows, "UC Desert Research and Extension Center Celebrates 100 Years."

9. Stevens, *Hoover Dam*, 9.

10. Hendricks, "Developing San Diego's Desert Empire."

11. "Death of Dr. Wozencraft," *Daily Evening Bulletin* (San Francisco), November 24, 1887.

12. Hays, *Conservation and the Gospel of Efficiency*, 243.

13. California State Water Resources Control Board, "Imperial Irrigation District Alleged Waste and Unreasonable Use of Water, Water Rights Decision 1600," 4.

14. Gottlieb and FitzSimmons, *Thirst for Growth*, 78–79.

15. Erie, *Beyond* Chinatown, 181.

16. California State Water Resources Control Board, "Imperial Irrigation District Alleged Waste," 58.

17. Gottlieb and FitzSimmons, *Thirst for Growth*, 81.

18. Ibid.

19. Boyarsky, "Imperial Valley Farmers Fear MWD Has Its Eye on Their Water."

20. Reisner and Bates, *Overtapped Oasis*, 152–58; Imperial Irrigation District, "Water Conservation Agreement Between Imperial Irrigation District and The Metropolitan Water District of Southern California."

21. Tina Shields, author interview, September 3, 2015.

22. Cohen, "Hazard's Toll," iv.

23. Petition of Imperial Irrigation District for Modification of Revised Water Rights Order 2002-0013, California State Water Resources Control Board, November 18, 2014.

24. Shields, "Crossroads at the Salton Sea."

25. US Bureau of Economic Analysis, "Regional Economic Accounts, Farm Income and Expenses" (table CA45).

Chapter 10 Notes

1. Terry Fulp, author interview, February 24, 2015.

2. Danielson, "Water Administration in Colorado—Higher-ority or Priority," 298.

3. US Bureau of Reclamation, "Lake Mead High and Low Elevations (1935–2014)," 1.

4. Scott Balcomb to Herb Guenther, personal correspondence, October 7, 2004.

5. Schiffer, Guenther, and Carr, "From a Colorado River Compact Challenge to the Next Era of Cooperation among the Seven Basin States," 217.

6. Fleck, "Abandoned Marina a Sign of Major Drought."

7. US Bureau of Reclamation, "Colorado River Basin Consumptive Uses and Losses Report 2001–2005," Table UC-9, 32.

8. Verburg, *The Colorado River Documents 2008*, Tables 2-1 through 2-5.

9. Boxall, "Running on Empty."

10. Statement of Leslie James (executive director, Colorado River Energy Distributors Association), "Opportunities and Challenges on Enhancing Federal Power Generation and Transmission," hearing before the House Subcommittee on Water and Power, Committee on Resources, 109th Congress (2005).

11. Ostler, "Upper Colorado River Basin Perspectives on the Drought," 18, 29.

12. Scott Balcomb to Herb Guenther, October 7, 2004.

13. John Entsminger, author interview, October 19, 2010.

14. Anne Castle, author interview, June 18, 2015.

15. Ostrom, "Social Capital: A Fad or a Fundamental Concept," 176.

16. John Entsminger, author interview, October 19, 2010.

17. Terry Fulp, author interview, April 23, 2010.

18. John Entsminger, author interview, October 19, 2010; Tom McCann, author interview, February 20, 2015.

19. McKinnon, "Arizona Fights Changes in Colorado River Plan."

20. John Entsminger, author interview, October 19, 2010.

21. David Donnelly, author interview, October 12, 2015; Terry Fulp, author interview, April 23, 2010; John Entsminger, author interview, October 19, 2010.

22. Fleck, "Arizona Water Managers Warn Lake Mead Could Be Sorta Unusable in Five to Eight Years."

Chapter 11 Notes

1. Darryl Vigil, statement to US Senate Energy and Natural Resources Committee, July 16, 2013.

2. Fleck, "Whitehorse Lake Sees Flowing Water at Last."

3. US Census Bureau, "American Community Survey"; Navajo Nation Department of Water Resources, "Draft Water Resource Development Strategy for the Navajo Nation," 49–53.

4. Western Regional Climate Center, "Cooperative Climatological Data Summaries."

5. Wilkinson, *Crossing the Next Meridian*, 219.

6. Chee Smith Jr., author interview, January 2014.

7. *Winters v. United States*, 207 U.S. 564, 28 S. Ct. 207, 52 L. Ed. 340 (1908).

8. Amy Cordalis, Tribal Reserved Rights and Settlements in the CRB, Clyde Martz Summer Water Conference, Boulder, CO, June 5, 2013.

9. Rifkind, Special Master's Report, 262.

10. Ibid., 263.

11. Price and Weatherford, "Indian Water Rights in Theory and Practice," 103.

12. Pollack, "*Navajo Nation v. Department of the Interior*," presentation, Continuing Legal Education Law of the River Conference, Las Vegas, NV, May 1, 2015.

13. US Department of Agriculture, "National Agricultural Statistics Service CropScape Cropland Data Layer."

14. Cordalis and Cordalis, "Indian Water Rights," 333; Kuhn, "Managing the Uncertainties of the Colorado River System," 23.

15. Fleck, "Navajos Stand to Gain Water Windfall."

16. Verberg, *The Colorado River Documents*, 5–26.

17. King et al., "Getting to the Right Side of the River," 78.

18. Dibble, "Calderón Stands Firm against Lining the All-American Canal."

19. King et al., "Getting to the Right Side of the River," 79.

20. Pacific Institute comments on Colorado River Interim Surplus Criteria Draft Environmental Impact Statement, included in: US Bureau of Reclamation, "Colorado River Interim Surplus Critieria Final Environmental Impact Statement," B-42.

21. Schlimgen-Wilson et al. to David Hayes, February 15, 2000, Attachment G in US Bureau of Reclamation, "Colorado River Interim Surplus Critieria Final Environmental Impact Statement."

22. Michael Cohen, author interview, August 19, 2015.

23. US Bureau of Reclamation, Colorado River Interim Surplus Criteria Final Environmental Impact Statement, 2–3.

24. "Environmental Suit Filed on Colorado River Plan," *New York Times*, June 30, 2000.

25. Jones et al., "Valuation in the Anthropocene," 1–47.

Chapter 12 Notes

1. MacDougal, "The Delta of the Rio Colorado," 10.
2. Leopold, *Sand County Almanac*, 141–42.
3. Mueller and Marsh, "Lost, A Desert River and Its Native Fishes," 2.
4. Ibid.
5. Cohen et al., "Water to Supply the Land," 55; Brun et al., "Agricultural Value Chains in the Mexicali Valley of Mexico," 5.
6. Brun et al., "Agricultural Value Chains in the Mexicali Valley of Mexico," 6.
7. Osvel Hinojosa-Huerta, author interview, February 10, 2015.
8. Peter Culp, author interview, March 23, 2014.
9. McKinnon, "New Yuma Reservoir Is a Water Saver."
10. US Bureau of Reclamation, "Colorado River Accounting and Water Use Report Arizona, California, and Nevada Calendar Year 2009," 23; Kara Gillon and Defenders of Wildlife, comments on Drop 2 Reservoir Environmental Impact Assessment.
11. US Bureau of Reclamation, "Name Change Approved for Drop 2 Storage Reservoir."
12. John Entsminger, author interview, October 19, 2010.
13. US Department of the Interior, "Secretary Kempthorne Announces Joint US-Mexico Statement on Lower Colorado River Issues," August 13, 2007.
14. Fradkin, *A River No More*, 319–41.
15. International Boundary and Water Commission, Minute No. 306.
16. Culp, "Minute 319 Negotiations."
17. Jenkins, "New Hope for the Delta."
18. Ibid.
19. Jennifer Pitt, author interview, September 2014.
20. Garcia et al., "Irrigation Engineering in Seismic Zones, Mexicali Valley, Mexico"; US Geological Survey "Earthquake Summary, Magnitude 7.2—Baja California, Mexico."
21. Sanchez and Cortez-Lara, "Minute 319 of the International Boundary and Water Commission between the US and Mexico."
22. US Department of Interior, "Salazar, Elvira Announce Water Agreement

to Support Response to Mexicali Valley Earthquake."

23. King et al., "Getting to the Right Side of the River," 90.

24. Terry Fulp, author interview, August 21, 2015.

25. The most detailed publicly available account of the final steps of the negotiations is Matt Jenkins's article "New Hope for the Delta," *High Country News*, January 17, 2014; see also International Boundary and Water Commission, Minute 319.

26. John Entsminger, author interview, February 25, 2015.

27. Bob Snow, personal communication, December 5, 2015.

Chapter 13 Notes

1. Statements of Senator Jeff Flake and Deputy Interior Secretary Michael Connor, Hearing on S. 1894, *California Emergency Drought Relief Act*, before the Senate Committee on Energy and Natural Resources (2015).

2. Phillips et al., *Reining in the Rio Grande*, 141.

3. Statement of John H. Bliss, Upper Colorado River Commissioner for State of New Mexico, Hearings before the House Subcommittee on Irrigation and Reclamation on H.R. 2494 and H.R. 2352, *San Juan–Chama Reclamation Project, and Navajo Indian Irrigation Project*; Committee on Interior and Insular Affairs, 86th Congress (1960).

4. Falk et al., *Estimated 2008 Groundwater Potentiometric Surface*, 2011, 1.

5. Oswald, "How Much Is Abiquiu Lake's Desert Shoreline Worth?"; Stomp, "Water Resources Management Strategy Update."

6. Sorensen, "Water Resources Drought Update"; Fleck, "Phoenix, Lake Mead, and 'the Anticommons'"; Fleck, "Priority Administration and Arizona's Colorado River Allotment."

7. Getches, "Competing Demands for the Colorado River," 420; Colorado River Governance Initiative, "Rethinking the Future of the Colorado River," 13.

8. This chapter owes significantly to ideas sketched out by Brad Udall, then at the University of Colorado, in a series of talks in the summer of 2013.

9. Fleck, "Brad Udall on the Colorado River and 'the Reality of the Public.'"

10. Michael Connor, author interview, March 28, 2014.

11. Herbert, "Southland's Water Safety Margin Place at 10 Years."

Bibliography

Books

August, Jack L. *Vision in the Desert: Carl Hayden and Hydropolitics in the American Southwest.* Fort Worth, TX: Texas Christian University Press, 1999.

Beaver, Deanna, and the Parker Area Historical Society. *Images of America: Parker.* Mount Pleasant, SC: Arcadia Publishing Company, 2008.

Cannavò, Peter F. *The Working Landscape: Founding, Preservation, and the Politics of Place.* Cambridge, MA: MIT Press, 2007.

Chibnik, Michael. *Anthropology, Economics, and Choice.* Austin, TX: University of Texas Press, 2011.

Clifford, Josephine. *Overland Tales.* Philadelphia: Claxton, Remsen, & Haffelfinger, 1877.

Cohen, Michael J., Juliet Christian-Smith, and John Berggren. *Water to Supply the Land: Irrigated Agriculture in the Colorado River Basin.* Boulder, CO: Pacific Institute, 2013.

Colby, Bonnie G., John E. Thorson, and Sarah Britton. *Negotiating Tribal Water Rights: Fulfilling Promises in the Arid West.* Tucson, AZ: University of Arizona Press, 2005.

Davis, Mike. *Dead Cities: And Other Tales.* New York: New Press, 2002.

Erie, Steven P. *Beyond* Chinatown: *The Metropolitan Water District, Growth, and the Environment in Southern California.* Palo Alto, CA: Stanford University Press, 2006.

Fradkin, Philip L. *A River No More: The Colorado River and the West.* Berkeley, CA: University of California Press, 1981.

Garfin, Gregg, and National Climate Assessment (US). *Assessment of Climate Change in Southwest United States: A Report Prepared for the National Climate Assessment.* National Climate Assessment Regional Technical Input Report Series. Washington, DC: Island Press, 2013.

Garrick, Dustin Evan. *Water Allocation in Rivers under Pressure: Water Trading, Transaction Costs and Transboundary Governance in the Western US and Australia.* Northampton, MA: Edward Elgar Publishing, 2015.

Glaeser, Edward L. *Triumph of the City: How Our Greatest Invention Makes Us Richer, Smarter, Greener, Healthier, and Happier.* New York: Penguin Press, 2011.

Glennon, Robert. *Unquenchable: America's Water Crisis and What to Do About It.* Washington, DC: Island Press, 2010.

Gottlieb, Robert, and Margaret FitzSimmons. *Thirst for Growth: Water Agencies as Hidden Government in California.* Tucson, AZ: University of Arizona Press, 1991.

Green, Dorothy. *Managing Water: Avoiding Crisis in California.* Berkeley, CA: University of California Press, 2007.

Hays, Samuel P. *Conservation and the Gospel of Efficiency: The Progressive Conservation Movement, 1890–1920.* Cambridge, MA: Harvard University Press, 1959.

Hundley, Norris, Jr. *Dividing the Waters: A Century of Controversy between the United States and Mexico.* Berkeley, CA: University of California Press, 1966.

Hundley, Norris, Jr. *The Great Thirst: Californians and Water, a History.* 2nd ed. Berkeley, CA: University of California Press, 2001.

Hundley, Norris, Jr. *Water and the West: The Colorado River Compact and the Politics of Water in the American West.* 2nd ed. Berkeley, CA: University of California Press, 2009.

Kupel, Douglas E. *Fuel for Growth.* Tucson, AZ: University of Arizona Press, 2003.

Lankford, Bruce A. *Resource Efficiency Complexity and the Commons: The Para-*

commons and Paradoxes of Natural Resource Losses, Wastes, and Wastages. London: Routledge, Taylor and Francis Group, 2013.

Leopold, Aldo. *A Sand County Almanac, and Sketches Here and There.* New York: Oxford University Press USA, 1989.

Mann, Dean E. *The Politics of Water in Arizona.* Tucson, AZ: University of Arizona Press, 1963.

Mendenhall, Walter C. *Development of Underground Waters in the Western Coastal Plain Region of Southern California.* Washington, DC: Government Printing Office, 1905.

Moehring, Eugene P. *Resort City in the Sunbelt.* Reno, NV: University of Nevada Press, 2000.

Moehring, Eugene P., and Michael S. Green. *Las Vegas: A Centennial History.* Reno, NV: University of Nevada Press, 2005.

Ostrom, Elinor. *Governing the Commons: The Evolution of Institutions for Collective Action.* Cambridge, UK: Cambridge University Press, 1990.

Ostrom, Elinor. "Public Entrepreneurship: A Case Study in Ground Water Basin Management." PhD diss., University of California, Los Angeles, 1964.

Ostrom, Elinor, Roy Gardner, and James Walker. *Rules, Games, and Common-Pool Resources.* Ann Arbor, MI: University of Michigan Press, 1994.

Ostrom, Vincent. *Water and Politics: A Study of Water Policies and Administration in the Development of Los Angeles.* Los Angeles: Haynes Foundation, 1953.

Phillips, Fred M., G. Emlen Hall, and Mary E. Black. *Reining in the Rio Grande: People, Land, and Water.* Albuquerque, NM: University of New Mexico Press, 2011.

Powell, James L. *Dead Pool: Lake Powell, Global Warming, and the Future of Water in the West.* Berkeley, CA: University of California Press, 2008.

Reisner, Marc. *Cadillac Desert: The American West and Its Disappearing Water.* New York: Viking, 1986.

Reisner, Marc, and Sarah F. Bates. *Overtapped Oasis: Reform or Revolution for Western Water.* Washington, DC: Island Press, 1990.

Sauder, Robert A. *The Yuma Reclamation Project: Irrigation, Indian Allotment, and Settlement Along the Lower Colorado River.* Reno: University of Nevada Press, 2009.

Stevens, Joseph E. *Hoover Dam: An American Adventure.* Norman, OK: University of Oklahoma Press, 1990.

Summitt, A. R. *Contested Waters: An Environmental History of the Colorado River.* Boulder, CO: University Press of Colorado, 2013.

Sykes, Godfrey Glenton. *The Colorado Delta.* Carnegie Institution of Washington. Publication no. 460; American Geographical Society Special publication, special publication no. 19. Washington, DC: Carnegie Institution of Washington and the American Geographical Society of New York, 1937.

Tarlock, A. Dan, James N. Corbridge, David H. Getches, Reed D. Benson, and Sarah F. Bates. *Water Resource Management: A Casebook in Law and Public Policy.* New York: Foundation Press, 2014.

Thorson, John E., Sarah Britton, and Bonnie G. Colby. *Tribal Water Rights: Essays in Contemporary Law, Policy, and Economics.* Tucson, AZ: University of Arizona Press, 2006.

Tyler, Daniel. *Silver Fox of the Rockies: Delphus E. Carpenter and Western Water Compacts.* Norman, OK: University of Oklahoma Press, 2003.

US Bureau of Reclamation. *Reclamation: Managing the Water in the West—The Bureau of Reclamation: History Essays from the Centennial Symposium.* 2 vols. Washington, DC: US Department of the Interior, 2008.

Verburg, Katherine O. *The Colorado River Documents, 2008.* Denver, CO: US Department of the Interior, Bureau of Reclamation, Lower Colorado Region, 2010.

Vollmann, Willliam T. *Imperial.* New York: Viking Press, 2009.

Walker, B. H., and David Salt. *Resilience Practice: Building Capacity to Absorb Disturbance and Maintain Function.* Washington, DC: Island Press, 2012.

Ward, Evan. *Border Oasis: Water and the Political Ecology of the Colorado River Delta, 1940–1975.* Tucson, AZ: University of Arizona Press, 2003.

Welsh, Frank. *How to Create a Water Crisis.* Boulder, CO: Johnson Publishing Company, 1985.

Wilkinson, Charles F. *Crossing the Next Meridian: Land, Water, and the Future of the West.* Washington, DC: Island Press, 1992.

Journal Articles, Conference Papers, Reports, Legal Cases, Legislation, etc.

Arizona v. California, 373 U.S. 546 (1963).

Arizona Interstate Stream Commission (now the Arizona Department of Water Resources). *Twenty-First Annual Report.* Phoenix, AZ, 1968.

Bali, Khaled. "Imperial Alfalfa Irrigation Requirement." Division of Agriculture and Natural Resources, University of California, 2016. http://ceimperial.ucanr.edu/Custom_Program275/Water_Quality_FAQs/.

Bell, Tina Marie. "Gila Project." Report. US Bureau of Reclamation, 1997.

Berman, Mindy. "A Tale of Two Transfers: Palo Verde and Imperial Valley Farmers Take Different Roads." *Aqueduct* 72 (2006): 3.

Blomquist, William. "Crafting Water Constitutions in California." Paper presented at "Vincent Ostrom: The Quest to Understand Human Affairs" conference, Bloomington, Indiana, May 31–June 3, 2006. http://www.indiana.edu/~voconf/papers/blomquist_voconf.pdf.

Boxall, Bettina. "Running on Empty." *Los Angeles Times*, October 17, 2004.

Boyarsky, Bill. "Imperial Valley Farmers Fear MWD Has Its Eye on Their Water." *Los Angeles Times*, October 3, 1983.

Brean, Henry. "Authority Approves Leasing Water to California." *Las Vegas Review-Journal*, September 17, 2015. http://www.reviewjournal.com/news/water-environment/authority-approves-leasing-water-california.

Brun, Lukas, Ajmal Abdulsamad, Christopher Geurtsen, and Gary Gereffi. "Agricultural Value Chains in the Mexicali Valley of Mexico." Report. Center on Globalization Governance & Competitiveness, 2010.

Cahill, Ryan, and Jay Lund. "Residential Water Conservation in Australia and California." *Journal of Water Resources Planning and Management* 139, no. 1 (2013): 117–21.

California Department of Food and Agriculture. "Dairy Production Data." Report. June 2015.

California Department of Water Resources. "California State Water Project at a Glance." Brochure. 2011.

California Emergency Drought Relief Act, Hearing on S. 1894, before the Senate Committee on Energy and Natural Resources (2015).

California Department of Water Resources. *California State Water Project at a Glance.* Sacramento, CA, April 2011. http://www.water.ca.gov/recreation/brochures/pdf/swp_glance.pdf.

California State Water Resources Control Board. "Urban Water Supplier." Report Dataset. Sacramento, CA, 2014–2016.

California State Water Resources Control Board. "Imperial Irrigation District Alleged Waste and Unreasonable Use of Water, Water Rights Decision 1600." Report. 1984.

California State Water Resources Control Board. "Petition of Imperial Irrigation District for Modification of Revised Water Rights Order 2002-0013." Sacramento, CA, November 18, 2014.

Campbell, George W. "Pinal County Agriculture." PhD diss., University of Arizona, 1959.

Carrillo-Guerrero, Yamilette Karina. "Water Conservation, Wetland Restoration, and Agriculture in the Colorado River Delta, Mexico." PhD diss., University of Arizona, 2009.

Carrillo-Guerrero, Yamilette K., Karl Flessa, Osvel Hinojosa-Huerta, and Laura López-Hoffman. "From Accident to Management: The Cienega de Santa Clara Ecosystem." *Ecological Engineering* 59 (2013): 84–92.

Coates, James. "Colorado Offers Water to California." *Chicago Tribune*, March 3, 1991.

Cohen, Michael. "Hazard's Toll: The Costs of Inaction at the Salton Sea." Report. Boulder, CO: Pacific Institute, 2014.

Cohen, Michael J., Jennifer C. Martin, Nancy Ross, and Paula Luu. "Municipal Deliveries of Colorado River Basin Water." Report. Oakland, CA: Pacific Institute, 2011.

Colorado River Commission, Minutes of the Twelfth Meeting, Santa Fe, NM, November 12, 1922. http://wwa.colorado.edu/resources/colorado-river/compact.html.

Colorado River Governance Initiative. "Rethinking the Future of the Colorado River: Draft Interim Report of the Colorado River Governance Initiative." 2010.

Connall, Desmond D., Jr. "A History of the Arizona Groundwater Management Act." *Arizona State Law Journal* no. 2 (1982): 313–40.

Cordalis, Amy, and Daniel Cordalis. "Indian Water Rights: How *Arizona v. California* Left an Unwanted Cloud over the Colorado River Basin." *Arizona Journal of Environmental Law and Policy* 5 (2014): 333–62.

Culp, Peter. "Minute 319 Negotiations." Lecture. Tamarisk Coalition Research and Management Conference, Albuquerque, New Mexico, February 10, 2015.

Daily Evening Bulletin (San Francisco). "Death of Dr. Wozencraft." November 24, 1887.

Danielson, Philip A. "Water Administration in Colorado—Higher-ority or Priority." *Rocky Mountain Law Review* 30 (1957): 293–314.

Davis, Tony. "Is California Trying to Take Our Water?" *Arizona Daily Star* (Tucson, AZ), June 27, 2015.

Deverell, William, and Tom Sitton. "Forget It, Jake." *Boom: A Journal of California* 3, no. 3 (2013): 3–7.

Dibble, Sandra. "Calderón Stands Firm against Lining the All-American Canal." *San Diego Union-Tribune*, May 5, 2007.

Dozier, Larry W., and Thomas W. McCann. "CAP Priority to Colorado River Water." Central Arizona Water Conservation District, Phoenix, AZ, undated.

Dumke, Glenn S. "Mission Station to Mining Town: Early Las Vegas." *Pacific Historical Review* (1953): 257–70.

Falk, Sarah E., Laura M. Bexfield, and Scott K. Anderholm. "Estimated 2008 Groundwater Potentiometric Surface and Predevelopment to 2008 Water-Level Change in the Santa Fe Group Aquifer System in the Albuquerque Area, Central New Mexico." Report. US Department of the Interior, US Geological Survey, 2011.

Fleck, John. "Abandoned Marina a Sign of Major Drought." *Albuquerque Journal*, July 19, 2009.

Fleck, John. "Arizona Water Managers Warn Lake Mead Could Be Sorta Unusable in Five to Eight Years." *Inkstain* (website), June 9, 2014. http://www.inkstain.net/fleck/2014/06/arizona-water-managers-warn-lake-mead-could-be-sorta-unusable-in-five-to-eight-years/.

Fleck, John. "Brad Udall on the Colorado River and 'the Reality of the Public.'" *Inkstain* (website), August 18, 2013. http://www.inkstain.net/fleck/2013/08/brad-udall-on-the-colorado-river-and-the-reality-of-the-public/.

Fleck, John. "Farming Against the Odds on the Rio Grande," *Albuquerque Journal*, July 30, 2013.

Fleck, John. "Navajos Stand to Gain Water Windfall." *Albuquerque Journal*, March 26, 2009.

Fleck, John. "Phoenix, Lake Mead, and 'the Anticommons.'" *Inkstain* (website), October 18, 2014. http://www.inkstain.net/fleck/2014/10/phoenix-lake-mead-and-the-anticommons/.

Fleck, John. "Priority Administration and Arizona's Colorado River Allotment." *Inkstain* (website), October 26, 2014. http://www.inkstain.net/fleck/2014/10/priority-administration-and-arizonas-colorado-river-allotment/.

Fleck, John. "State's Future Banks on Colorado River." *Albuquerque Journal*, March 17, 2009.

Fleck, John. "When Water Supplies Ebb, Users Go with the Flow." *Albuquerque Journal*, November 11, 2014.

Fleck, John. "Whitehorse Lake Sees Flowing Water at Last." *Albuquerque Journal*, January 5, 2014.

Forsythe, James L. "World Cotton Technology Since World War II." *Agricultural History* 54, 1 (1980): 208–22.

Garcia, Dennis, C. M. Burt, and Maria Paredes Vallejo. "Irrigation Engineering in Seismic Zones, Mexicali Valley, Mexico." Proceedings of the Sixth International Conference on Irrigation and Drainage, United States Committee on Irrigation and Drainage (USCID), San Diego, CA, November 15–18, 2011.

Gerlak, Andrea K. "Resistance and Reform: Transboundary Water Governance in the Colorado River Delta." *Review of Policy Research* 32, no. 1 (2015): 100–123.

Getches, David H. "Competing Demands for the Colorado River." *University of Colorado Law Review* 56 (1984): 413–79.

Gammage, Grady, and Morrison Institute for Public Policy. "Watering the Sun Corridor: Managing Choices in Arizona's Megapolitan Area." Phoenix, AZ: Morrison Institute for Public Policy, Arizona State University, 2011.

Gillon, Kara, and Defenders of Wildlife. Comments on Drop 2 Reservoir Environmental Impact Assessment. February 15, 2007. http://pacinst.org/wp-content/uploads/2013/02/drop_2_comments3.pdf.

Gleick, Peter H., and Meena Palaniappan. "Peak Water Limits to Freshwater Withdrawal and Use." *Proceedings of the National Academy of Sciences* 107.25 (2010): 11155–62.

Glennon, Robert, and Jennifer Pitt. "Our Water Future Needs Creativity." *Arizona Republic*, May 10, 2004.

Grant, Daniel. "Tapping the Past for California's Water Future." *Edge Effects* (website), June 23, 2015. http://edgeeffects.net/tapping-the-past/.

Grant, Stanley B., Tim D. Fletcher, David Feldman, Jean-Daniel Saphores, Perran L. M. Cook, Mike Stewardson, Kathleen Low, Kristal Burry, and Andrew J. Hamilton. "Adapting Urban Water Systems to a Changing Climate: Lessons from the Millennium Drought in Southeast Australia." *Environmental Science and Technology* 47, no. 19 (2013): 10727–34.

Griffin, Daniel, and Kevin J. Anchukaitis. "How Unusual Is the 2012–2014 California Drought?" *Geophysical Research Letters* (2014). doi:10.1002/2014GL062433.

Hardin, Garrett. "The Tragedy of the Commons." *Science* 162, no. 3859 (1968): 1243–48.

Harrison, Christian. "Water Use and Natural Limits in the Las Vegas Valley: A History of the Southern Nevada Water Authority." Masters thesis, University of Nevada Las Vegas, 2009.

Hendricks, William O. "Developing San Diego's Desert Empire." *Journal of San Diego History* 17, no. 3 (1971). http://www.sandiegohistory.org/jour nal/71summer/desert.htm.

Herbert, Ray. "Southland's Water Safety Margin Place at 10 Years." *Los Angeles Times*, January 24, 1960.

Hettena, Seth. "Interior Secretary Cuts California's Share of Colorado River Water." Associated Press. December 17, 2002.

Howitt, Richard, and Ellen Hanak. "Incremental Water Market Development: The California Water Sector 1985–2004." *Canadian Water Resources Journal* 30, no. 1 (2005): 73–82.

Imperial County (CA) Agricultural Commission. "Imperial County Agricultural Crop and Livestock Report," 2000, 2014.

Imperial Irrigation District. "Monthly Crop Acreage Report." Imperial, CA: Imperial Irrigation District, December 2015. http://www.iid.com/water /agriculture-customers/water-and-crop-news.

Imperial Irrigation District. "Water Conservation Agreement Between Imperial Irrigation District and The Metropolitan Water District of Southern California," CVC01.02. December 1989. http://www.iid.com/home /showdocument?id=9937.

Imperial Valley Press. "Get Ready for the Big Grab." February 25, 2007.

International Boundary and Water Commission, United States of America and Mexico. Minute No. 306. El Paso, TX, 2000. http://www.ibwc.state.gov /Treaties_Minutes/Minutes.html.

International Boundary and Water Commission, United States of America and Mexico. Minute No. 319. Coronado, CA, 2012. http://www.ibwc.state .gov/Treaties_Minutes/Minutes.html.

Jenkins, Matt. "New Hope for the Delta." *High Country News*, January 17, 2014.

Jerla, Carly Starr. "An Analysis of Coordinated Operation of Lakes Powell and Mead under Lower Reservoir Conditions." Masters thesis, University of Colorado, Boulder, 2005.

Jones, Benjamin A., Robert P. Berrens, Hank C. Jenkins-Smith, Carol L. Silva, Deven E. Carlson, Joseph T. Ripberger, and Kuhika Gupta. "Valuation in the Anthropocene." Working Paper, Center for Energy Security and Society, University of Oklahoma, 2015.

Judkins, Gabriel L., and Kelli Larson. "The Yuma Desalting Plant and Cienega de Santa Clara Dispute: A Case Study Review of a Workgroup Process." *Water Policy* 12, no. 3 (2010): 401–15.

Kightlinger, Jeffrey. "The Lower Colorado River Multi-Species Conservation Program." *Pacific McGeorge Global Business and Development Law Journal* 19 (2006): 33–46.

King, Jonathan S., Peter W. Culp, and Carlos de la Parra. "Getting to the Right Side of the River: Lessons for Binational Cooperation on the Road to Minute 319." *University of Denver Water Law Review* 18 (2014): 36–114.

Konikow, Leonard F. "Groundwater Depletion in the United States (1900–2008)." Report. Washington, DC: US Department of the Interior, US Geological Survey, 2013.

Kuhn, Eric. "The Colorado River: The Story of a Quest for Certainty on a Diminishing River." Report. Colorado River Water Conservation District, 2007.

Kuhn, Eric. "Managing the Uncertainties of the Colorado River System." In *How the West Was Warmed: Responding to Climate Change in the Rockies*, edited by Beth Conover, 100–110. Golden, CO: Fulcrum Publishing, 2009.

Las Vegas Age. "Action of 7 States Means Millions to Las Vegas." November 25, 1922.

Las Vegas Sun. "How Much Water Evaporates from the Bellagio Fountains?" April 14, 2010.

Las Vegas Valley Water Authority. "Consolidated Annual Financial Report." 2014.

Laverty, Finley B., and Herbert A. van der Goot. "Development of a Freshwater Barrier in Southern California for the Prevention of Sea Water Intrusion." *Journal of the American Water Works Association* 47, no. 9 (1955): 886–908.

Leavenworth, Stuart. "Colorado Conversion? A Key Interior Official Professes His Love for the Issue-Rich River." *Sacramento Bee*, May 30, 2004.

Lochhead, James S. "Upper Basin Perspective on California's Claims to Water from the Colorado River Part I: The Law of the River." *University of Denver Water Law Review* 4 (2000): 290–330.

Lochhead, James S. "Upper Basin Perspective on California's Claims to Water from the Colorado River Part II: The Development, Implementation and Collapse of California's Plan to Live within Its Basic Apportionment." *University of Denver Water Law Review* 6 (2002): 318–410.

Loomis, Brandon. "Ducey: Don't Punish Arizona for Its Water Conservation." *Arizona Republic*, June 9, 2015.

Los Angeles Department of Water and Power. "2005 Urban Water Management Plan." Report. Los Angeles, CA, 2006. http://www.waterboards .ca.gov/water_issues/programs/conservation_portal/conservation_report ing.shtml.

Los Angeles Times. "Dam Storm Thunders." May 23, 1928.

Los Angeles Times. "Ground Water: The Perils to Its Purity." December 12, 1989. http://articles.latimes.com/1989-12-10/news/ss-578_1_drinking-water.

Los Angeles Times. "Hawthorne Mayor Hits 'Double Tax.'" November 15, 1959.

Los Angeles Times. "Land, Water, Homes, Stability, and Progress." August 9, 1903.

Los Angeles Times. "Warning Given on Salt Barrier." July 26, 1946.

Lustgarten, Abrahm. "The 'Water Witch'—Pat Mulroy Preached Conservation while Backing Growth in Las Vegas." *ProPublica*, June 2, 2015. https://projects.propublica.org/killing-the-colorado/story/pat-mulroy-las -vegas-water-witch.

Martin, William E., and Robert A. Young. "The Need for Additional Water in the Arid Southwest: An Economist's Dissent." *Annals of Regional Science* 3, no. 1 (1969): 22–31.

MacDonnell, Lawrence J. "*Arizona v. California* Revisited." *Natural Resources Journal* 52 (2012): 363–420.

MacDougal, Daniel Trembly. "The Delta of the Rio Colorado." *Bulletin of the American Geographical Society* (1906): 1–16.

McKinnon, Shaun. "Arizona Fights Changes in Colorado River Plan." *Arizona Republic* (Phoenix, AZ), October 4, 2007.

McKinnon, Shaun. "New Yuma Reservoir Is a Water Saver." *Arizona Republic* (Phoenix, AZ), November 26, 2010.

Meadows, Robin. "Research News: UC Desert Research and Extension Center Celebrates 100 Years." *California Agriculture* 66, no. 4 (2012): 122–26.

Medellín-Azuara, Josué, Jay Lund, and Richard Howitt, "Jobs per Drop Irrigating California Crops." *California WaterBlog* (website), April 28, 2015. http://californiawaterblog.com/2015/04/28/jobs-per-drop-irrigating-california-crops/.

Meister, Herman S. "Sample Cost to Establish and Produce Wheat, Imperial County." Report. University of California, Davis. UC Cooperative Extension Service WH-IM-04-1. 2004.

Mendenhall, Walter C. *Development of Underground Waters in the Central Coastal Plain Region of Southern California.* Washington, DC: Government Printing Office, 1905.

Metropolitan Water District of Southern California. *Annual Progress Report to the California State Legislature.* Los Angeles, CA, 2004.

Metropolitan Water District of Southern California. "Southern California's Integrated Water Resources Plan." Report no. 1107. Los Angeles, CA, March 1996.

Mueller, Gordon A., and Paul C. Marsh. "Lost, a Desert River and Its Native Fishes: A Historical Perspective of the Lower Colorado River." Report no. USGS/BRD/ITR—2002—0010. U.S. Geological Survey, Fort Collins Science Center, 2002.

Mulroy, Patricia. "Beyond the Division: A Compact That Unites." *Journal of Land Resources & Environmental Law* 28 (2008): 105–17.

Nash, Linda L., and Peter H. Gleick. "The Colorado River Basin and Climatic Change: The Sensitivity of Streamflow and Water Supply to Variations in Temperature and Precipitation." Report no. EPA230-R-93-009. Washington, DC: US Environmental Protection Agency, 1993.

Navajo Nation Department of Water Resources. "Draft Water Resource Development Strategy for the Navajo Nation." Window Rock, AZ, July 2011.

Nelson, A. M. Development Agent, Imperial County. "Increase in Crop Acreages and Property Values in Imperial County." 1917. Collection of Imperial Irrigation District.

Nevada Gaming Control Board. "Nevada Gaming Abstract." Las Vegas, NV, 2014.

New York Times. "Environmental Suit Filed on Colorado River Plan." June 30, 2000.

New York Times. "Senators Battle Over Boulder Dam." May 30, 1928.

Ostler, Don. "Upper Colorado River Basin Perspectives on the Drought." *Southwest Hydrology* (March–April 2005): 18, 25.

Ostrom, Elinor. "Beyond Markets and States: Polycentric Governance of Complex Economic Systems." *American Economic Review* 100, no. 3 (2010): 641–72.

Ostrom, Elinor. "A Long Polycentric Journey." *Annual Review of Political Science* 13 (2010): 1–23.

Ostrom, Elinor. "Social Capital: A Fad or a Fundamental Concept?" *Social Capital: A Multifaceted Perspective* 172, no. 173 (2000): 195–98.

Ostrom, Elinor. "Why Do We Need to Protect Institutional Diversity?" *European Political Science* 11, no. 1 (2012): 128–47.

Oswald, Mark. "How Much Is Abiquiu Lake's Desert Shoreline Worth?" *Albuquerque Journal*, April 28, 2014.

Pacific Institute. "Surplus Criteria Proposal by Pacific Institute," Attachment G, February 15, 2000, in US Bureau of Reclamation, "Colorado River Interim Surplus Critieria Final Environmental Impact Statement." Washington, DC: US Department of the Interior, December 2000.

Pitt, Jennifer, Daniel F. Luecke, Michael J. Cohen, and Edward P. Glenn. "Two Nations, One River: Managing Ecosystem Conservation in the Colorado River Delta." *Natural Resources Journal* 40 (2000): 819–64.

Pollack, Stanley. "*Navajo Nation v. Department of the Interior.*" Presentation at the Continuing Legal Education Law of the Colorado River Conference, Las Vegas, NV, May 1, 2015.

Poulsen, Brian D., Jr. "Reduce? A Look at the Upper Colorado River Basin's Annual Delivery Obligation to the Lower Basin in Light of Secretary Norton's Mid-Year 2005 AOP Decision." *Journal of Land Resources and Environmental Law* 26 (2005): 195–206.

Price, Monroe E., and Gary D. Weatherford. "Indian Water Rights in Theory and Practice: Navajo Experience in the Colorado River Basin." *Law and Contemporary Problems* 40 (Winter 1976): 97–131.

Putnam, Dan, Joe Brummer, Dennis Cash, Alan Gray, Tom Griggs, Mike Ottman, Ian Ray, Willie Riggs, Mark Smith, Glenn Shewmaker, and Rodney Todd. "The Importance of Western Alfalfa Production." Proceedings of 29th National Alfalfa Symposium and 30th California Alfalfa Symposium, Las Vegas, NV, 2000.

Rifkind, Simon. "Report of the Special Master—*Arizona v. California.*" Washington, DC: US Supreme Court, 1960.

Rittel, Horst W. J., and Melvin M. Webber. "Dilemmas in a General Theory of Planning." *Policy Sciences* 4, no. 2 (1973): 155–69.

Russelle, Michael. "After an 8,000-Year Journey, the 'Queen of Forages' Stands Poised to Enjoy Renewed Popularity." *American Scientist* 89 (2001): 252–61.

Sabo, John L., Tushar Sinha, Laura C. Bowling, Gerrit H. W. Schoups, Wesley W. Wallender, Michael E. Campana, Keith A. Cherkauer, et al. "Reclaiming Freshwater Sustainability in the Cadillac Desert." *Proceedings of the National Academy of Sciences* 107, no. 50 (2010): 21263–69.

St. Louis Federal Reserve. "Resident Population in Arizona." https://research.stlouisfed.org/fred2/series/AZPOP.

Sanchez, Vicente, and Alfonso A. Cortez-Lara. "Minute 319 of the International Boundary and Water Commission between the US and Mexico: Colorado River Binational Water Management Implications." *International Journal of Water Resources Development* 31, no. 1 (2015): 17–27.

Sarewitz, Daniel. "How Science Makes Environmental Controversies Worse." *Environmental Science and Policy* 7, no. 5 (2004): 385–403.

Schiffer, W. Patrick, Herbert R. Guenther, and Thomas G. Carr. "From a Colorado River Compact Challenge to the Next Era of Cooperation among the Seven Basin States." *Arizona Law Review* 49 (2007): 217–33.

Schuster, Elizabeth, and Bonnie Colby. "Farm and Ecological Resilience to Water Supply Variability." *Journal of Contemporary Water Research and Education* 151, no. 1 (2013): 70–83.

Semitropic Water Storage District. Combined Financial Statements. Wasco, CA, 2013. http://www.semitropic.com/FinancialStatements.htm.

Shields, Tina. "Crossroads at the Salton Sea." Presentation at the Continuing Legal Education Law of the River conference, Las Vegas, NV, May 1, 2015.

Snape, William, III, Michael Senatore, Kara Gillon, and Susan George. "Protecting Ecosystems under the Endangered Species Act: The Sonoran Desert Example." *Washburn Law Journal* 41 (2001).

Sorensen, Kathryn. "Water Resources Drought Update." Report to Phoenix, AZ, City Council, October 21, 2014.

Southern Nevada Water Authority. "Member Agency Estimated Population and Annual Water Usage 1994–2013." Dataset provided by SNWA, June 2014.

Southern Nevada Water Authority. "Water Resource Plan 09." Report. Las Vegas, NV: Southern Nevada Water Authority, 2009.

Southern Nevada Water Authority. "Water Conservation Plan 2014–2018." Report. Las Vegas, NV: Southern Nevada Water Autority, 2014.

Southern Nevada Water Authority. "Water Resource Plan 2015." Report. Las Vegas, NV: Southern Nevada Water Authority, 2015.

Stomp, John. "Water Resources Management Strategy Update." Presentation at the Albuquerque Bernalillo County Water Utility Authority board meeting, Albuquerque, NM, September 23, 2015.

US Bureau of Economic Analysis. Regional Economic Accounts. "Annual Gross Domestic Product (GDP) by State." http://www.bea.gov/regional/index.htm.

US Bureau of Economic Analysis. "Regional Economic Accounts, Farm Income and Expenses" (table CA45). http://www.bea.gov/regional/index.htm.

US Bureau of Reclamation. "Agreement Requesting Apportionment of California's Share of the Waters of the Colorado River Among the Applicants in the State" (commonly known as the "The California Seven Party Water Agreement of 1931"). Washington, DC: US Department of the Interior, August 18, 1931. http://www.usbr.gov/lc/region/pao/pdfiles/ca7pty.pdf.

US Bureau of Reclamation. "Colorado River Accounting and Water Use Report—Arizona, California, and Nevada." 2003–2014; "Compilation of Records in Accordance with Article V of the Decree of the Supreme Court of the United States in *Arizona v. California.*" Washington, DC: US Department of the Interior, 1964–2002.

US Bureau of Reclamation. "Colorado River Basin Consumptive Uses and Losses Report." Washington, DC: US Department of the Interior, 1971–2015. http://www.usbr.gov/uc/library/envdocs/reports/crs/crsul.html.

US Bureau of Reclamation. "Colorado River Basin Natural Flow and Salt Data." Washington, DC: US Department of the Interior. http://www.usbr.gov/lc/region/g4000/NaturalFlow/current.html.

US Bureau of Reclamation. "Colorado River Basin Water Supply and Demand Study." Report no. SR-26. Washington, DC: US Department of the Interior, 2012.

US Bureau of Reclamation. "Colorado River Interim Surplus Criteria Final Environmental Impact Statement." Report. Washington, DC: US Department of the Interior, 2000.

US Bureau of Reclamation. "Compilation of Records in Accordance with Article V of the Decree of the Supreme Court of the United States in *Arizona v. California.*" Washington, DC: US Department of the Interior, 1987, 1993.

US Bureau of Reclamation. "Groundwater Banking Pilot Project of Central Valley Project Water from City of Tracy to Semitropic Water Storage District." Report no. EA-05-111. Washington, DC: US Department of the Interior, October 2006.

US Bureau of Reclamation. "Lake Mead High and Low Elevations (1935–2015)." Washington, DC: US Department of the Interior, 2015. http://www.usbr.gov/lc/region/g4000/lakemead_line.pdf.

US Bureau of Reclamation. "Lower Colorado River Operations, Lake Mead at Hoover Dam." Washington, DC: US Department of the Interior, 2016. http://www.usbr.gov/lc/region/g4000/hourly/mead-elv.html.

US Bureau of Reclamation, "Moving Forward: Phase 1 Report." Washington, DC: US Department of the Interior, 2015.

US Bureau of Reclamation. "Name Change Approved for Drop 2 Storage Reservoir." Report. Washington, DC: US Department of the Interior, September 15, 2010.

US Census Bureau. "American Community Survey." Washington, DC: US Department of Commerce, 2014. https://www.census.gov/programs-surveys/acs/.

US Census Bureau. "Sixteenth Census of the United States: 1940." Washington, DC: US Department of Commerce, 1940.

US Census Bureau. "2014 Population Estimates, QuickFacts." Washington, DC: US Department of Commerce, 2014. http://www.census.gov/quickfacts/table/PST045215/32003,00.

US Department of Agriculture. "Census of Agriculture, 1978." Washington, DC: Government Printing Office, 1978.

US Department of Agriculture. "Census of Agriculture, 2012." Washington, DC: Government Printing Office, 2012.

US Department of Agriculture. "Census of Agriculture, 2014." Washington, DC: Government Printing Office, 2014.

US Department of Agriculture. "Crop Production—2014 Summary." Report. Washington, DC: Government Printing Office, 2015.

US Department of Agriculture. "Fresh Fruit and Vegetable Shipments, 2014." Report. Washington, DC: Government Printing Office, 2014.

US Department of Agriculture. "Milk Production." Report. Washington, DC: Government Printing Office, August 2015.

US Department of Agriculture, National Agricultural Statistics Service. "CropScape Cropland Data Layer." Washington, DC, 2015. https://nassgeodata.gmu.edu/CropScape/.

US Department of the Interior. Press release. "Salazar, Elvira Announce Water Agreement to Support Response to Mexicali Valley Earthquake." Washington, DC, December 20, 2010.

US Department of the Interior. Press release. "Secretary Kempthorne Announces Joint US-Mexico Statement on Lower Colorado River Issues." Washington, DC, August 13, 2007.

US Geological Survey. "Earthquake Summary, Magnitude 7.2—Baja California, Mexico." Report. Washington, DC, April 2010.

US Geological Survey. "Estimated Use of Water in the United States." Washington, DC: US Department of the Interior, 1950–2010.

US Geological Survey. "Largest Rivers in the United States." Open-File Report 87-242. Washington, DC: US Department of the Interior, 1990.

US House of Representatives, Subcommittee on Water and Power, Committee on Resources. H.R. Rep. No. 107-78 at 43 (2001).

US House of Representatives, Subcommittee on Water and Power, Committee on Resources. Oversight Hearing on Collaboration on the Colorado River, April 9, 2010.

US House of Representatives, Subcommittee on Water and Power, Committee on Resources, 109th Congress, Hearing on Opportunities and Challenges on Enhancing Federal Power Generation and Transmission. February 10, 2005. Washington, DC: US Government Printing Office, 2005.

US Senate, Energy and Natural Resources Committee, 113th Congress, first session. Hearing to review water resource issues in the Colorado River Basin, July 16, 2013. Washington, DC: US Government Printing Office, 2013.

US Senate, Subcommittee of the Committee on Irrigation and Reclamation, 78th Congress, second session. Hearings on Arizona water resources, July 31 and August 1–4, 1944. Washington, DC: US Government Printing Office, 1944.

Vano, Julie A., Tapash Das, and Dennis P. Lettenmaier. "Hydrologic Sensitivities of Colorado River Runoff to Changes in Precipitation and Temperature." *Journal of Hydrometeorology* 13, no. 3 (2012): 932–49.

Vano, Julie A., Bradley Udall, Daniel R. Cayan, Jonathan T. Overpeck, Levi D. Brekke, Tapash Das, Holly C. Hartmann, et al. "Understanding Uncertainties in Future Colorado River Streamflow." *Bulletin of the American Meteorological Society* 95, no. 1 (2014): 59–78.

Vegas Artesian Water Syndicate. Prospectus, 1905. Quoted at UNLV Libraries Digital Collections, Irrigation. http://digital.library.unlv.edu/collections /historic-landscape/irrigation.

Water Replenishment District of Southern California. "Engineering and Survey Report." Lakewood, CA: Water Replenishment District of California, 2015.

Webb, Richard M. T., and Darius J. Semmens. *Planning for an Uncertain Future—Monitoring, Integration, and Adaptation.* Report no. 2009-5049. Washington, DC: US Geological Survey, 2009.

Western Regional Climate Center. "Cooperative Climatological Data Summaries." Reno, NV. http://www.wrcc.dri.edu/climatedata/climsum/.

Western Regional Climate Center. Historic Climate Division data, South Coast Drainage Division. http://www.wrcc.dri.edu/cgi-bin/divplot1_form .pl?0406.

Woestendiek, John. "In Middle of Desert, Las Vegas Builds Lake. It May Run Out of Water, but Not Fantasies." *Philadelphia Inquirer*, June 15, 1992.

Wood, Daniel. "California Drought Springs New Limits on Developers." *Christian Science Monitor*, April 22, 1991.

Woodhouse, Connie A., Stephen T. Gray, and David M. Meko. "Updated Streamflow Reconstructions for the Upper Colorado River Basin." *Water Resources Research* 42, no. 5 (2006): 1–16.

Young, Robert A. "The Arizona Water Controversy: An Economist's View." *Journal of the Arizona Academy of Science* 6, no. 1 (1970): 3–10.

Yuma County Agriculture Water Coalition. "A Case Study in Efficiency— Agriculture and Water Use in the Yuma, Arizona, Area." Report. February 2015.

Yuma Desalting Plant / Cienega de Santa Clara Workgroup. "Balancing Water Needs on the Lower Colorado River." Report. April 22, 2005.

Yuma (Arizona) *Sun*. "Sen. Hayden Takes Step in Billion Dollar Dream." June 9, 1963.

Zetland, David. "The End of Abundance: How Water Bureaucrats Created and Destroyed the Southern California Oasis." *Water Alternatives* 2, no. 3 (2009): 350–69.

Zetland, David. "Will the People of Imperial Valley Jump or Get Pushed?" *Aguanomics* (website), November 11, 2014. http://www.aguanomics .com/2014/11/will-people-of-imperial-valley-jump-or.html.

Index